ONE RENEGADE CELL

How Cancer Begins

ROBERT A. WEINBERG

BASIC
BOOKS

A Member of the
Perseus Books Group

The Science Masters Series is a global publishing venture consisting of original science books written by leading scientists and published by a worldwide team of twenty-six publishers assembled by John Brockman. The series was conceived by Anthony Cheetham of Orion Publishers and John Brockman of Brockman Inc., a New York literary agency, and developed in coordination with Basic Books.

••••••••••••••

The Science Masters name and marks are owned by and licensed to the publisher by Brockman Inc.

••••••••••••••

••••••••••••••

Published by Basic Books,
A Member of the Perseus Books Group

••••••••••••••

••••••••••••••

FIRST EDITION.

••••••••••••••

A CIP catalog record for this book is available from the Library of Congress.
ISBN 0–465–07275–5

••••••••••••••

98 99 00 01 ❖/RRD 10 9 8 7 6 5 4 3 2 1

CONTENTS

ACKNOWLEDGMENTS

The work of many colleagues, too numerous to mention, has provided me with the insights that made the writing of this book possible. William Frucht, Senior Editor of Perseus Books, helped me convert my original text into one far more readable. I am very grateful to him.

....

CHAPTER 1

THE ENEMY WITHIN: GENES, CELLS, AND THE NATURE OF CANCER

Cancer wreaks havoc in almost every part of the human body. Tumors strike the brain and the gut, muscles and bones. Some grow slowly; others are more aggressive and expand quickly. Their presence in human tissues signals chaos and a breakdown of normal function. Cancer brings unwelcome change to a biological machine that is perfect, marvelously beautiful, and complex beyond measure. Wherever tumors appear, they take on the appearance of alien life forms, invaders that enter the body through stealth and begin their programs of destruction from within. But appearances deceive: The truth is much more subtle and endlessly interesting.

Tumors are not foreign invaders. They arise from the same material used by the body to construct its own tissues. Tumors use the same components—human cells—to form the jumbled masses that disrupt biological order and function and, if left unchecked, to bring the whole complex, life-sustaining edifice that is the human body crashing down.

How are human tissues put together from single cells? The description above might suggest the involvement of master builders who oversee crews of workers, directing them in the detailed construction of normal and malignant tissues. In reality, there are no overseers forcing throngs of cells to line up and assemble themselves into normal or cancerous tissues. Architectural complexity in living tissue comes from the bricks themselves, the individual cells. Control is exercised from the bottom up.

Normal and malignant cells know how to build. Each carries its own agenda that tells it when it should grow and divide and how it should aggregate with other cells to create organs and tissues. Our bodies are nothing more than highly complex societies of rather autonomous cells, each retaining many of the attributes of a fully independent organism.

Right there, we confront great beauty and profound danger. The beauty lies in the coordinated behavior of so many cells to create the single, highly functional cooperative that is the human body. The danger lies in the absence of a single overseeing master builder, which seems to put the whole enterprise at great risk. Granting autonomy to trillions of worker cells invites chaos. When, as usually happens, these cells are well behaved and public-spirited, extraordinarily complex order ensues. But on occasion, a cell may choose to go its own way and invent its own novel version of a tissue or organ. It is then that we see the much-feared chaos that we call cancer.

Most human tumors comprise a billion or more cells before we become aware of them. The cells within a tumor differ from their normal counterparts in many respects, exhibiting distinctive shapes, growth properties, and metabolism. The sudden appearance of such a horde of cells would seem to reflect some recent mass conversion, in which millions of normal cells enlisted overnight in the ranks of a tumor mass.

Once again, appearances deceive. The creation of a tumor is an extraordinarily slow process, often extending over decades. The cells forming a tumor are all lineal descendants of a single progenitor, a distant ancestor that lived many years before the tumor mass became apparent. This founder, this renegade cell, decided to go off on its own, to begin its own growth program within one of the body's tissues. Thereafter, its proliferation was controlled by its own internal agenda rather than the needs of the community of cells around it.

So there were no millions of recruits, only a single one that spawned a vast horde of like-minded descendants. The billions of cells in a tumor are cast in the image of their renegade ancestor. They have no interest in the well-being of the tissue and organism around them. Like the founder cell, they have only one program in mind: more growth, more replicas of themselves, unlimited expansion.

The chaos they create makes it clear how very dangerous it is to entrust each cell in the human body with its own measure of independence. Still, that is how we are put together, and how all complex, many-celled organisms have been designed for the past 600 million years. Learning this, we realize that the chaos of cancer is not a modern affliction but a risk run by all multicellular organisms, from ancient to modern. Indeed, given the trillions of cells in the human body, is it not a wonder that cancer does not erupt often during our long lives?

THE INTERNAL BLUEPRINT
....

To understand how a tumor grows, we need to understand the cells that form it. What caused that single founding cell to run amok? More generally, how does any cell, normal or malignant, know when to grow? Do cells have their own minds? And if not, what complex decision-making apparatus inside the living human cell determines its growth, quiescence, or death?

Our focus in this book is on the internal program carried by each normal human cell that tells it how and when to grow and associate with other cells to create the highly functional communities that are human tissues. The programs carried by various cells represent complex biological scripts, blueprints for their behavior. As we shall see, it is this internal program that is altered when cancer begins. Only after we understand this program in its normal and damaged forms can we understand the engine that drives the cancer cell.

There are several hundred types of cells in the human body. These various cell types aggregate to form distinct tissues and organs. Knowing this variability of individual cells, we might imagine the existence of a correspondingly large number of distinct scripts, each carried by a different cell type, each dictating a distinct agenda of growth and tissue-forming abilities. Here our intuition leads us astray. In truth, the cells in different parts of our body—in the brain, muscles, liver, and kidneys—as different as they appear, are really very much alike and, unexpectedly, all carry the same blueprint.

This sameness can be traced to their common origin. Like the cells in a tumor, all the cells that constitute a normal body descend from a common ancestral cell. They are cousins, members of an extended family. Through repeated rounds of growth and division, the single cell that is the fertilized egg yields the trillions of cells that form a complete body. The number of cells in an adult human—more than ten thousand billion—vastly exceeds the mind's ability to grasp.

The blueprint that directs all cells in the body is present in the ancestral fertilized egg and is then passed on, virtually unaltered, to all the descendant cells throughout the body. These trillions of cells can look and act very differently from one another, yet they all carry the same set of instructions that program their behavior. So there is a striking discrepancy between the cells' common inner blueprint and their highly diverse outer appearances. It seems that appearance tells us little about the internal program that guides the lives of these cells.

How can a single, common blueprint generate such diversity? Over the past decades the answer has emerged, and it is a simple one: The complex master plan carried by all cells in the body bears far more information than any single cell can use. Each cell in the body consults its master blueprint selectively, reading out only certain information from the vast library that it carries. This information is then used to choreograph its behavior. Selective reading allows each cell to behave differently from its relatives, close and distant, throughout the body.

Shortly after an egg is fertilized, it divides, as do its two daughters in turn. The embryonic development that follows is a frenzy of cell growth and division. The descendants of a fertilized egg arising over the next several cell generations seem very much alike; they stick together tightly, forming a homogeneous, undifferentiated cluster of cells that resembles a tiny raspberry. As embryonic development proceeds, however, the progeny of these cells begin to show differences. They start to become members of the communities of muscle cells or brain cells or blood cells. This process of choosing distinct fates—the process of differentiation—is the central mystery of human development and the obsession of those who study it.

A cell in one corner of the embryo reads out genetic instructions on how to make hemoglobin and becomes a red blood cell; a cell elsewhere consults information on the making of digestive enzymes and becomes part of the pancreas; a third cell reads the information on how to emit electrical signals and forms part of the brain.

But this decision by each embryonic cell to assume distinctive, differentiated traits, reached through selective reading of the genes that it carries, is not the only important decision the cell must make. It must also consult its genetic blueprint on another, equally weighty issue: when it should grow and divide, and when it should stop growing.

These instructions about growth remain important long after the embryonic stage. In most adult tissues, cells are continually dying and being replaced. Indeed, the ability of an adult tissue to maintain its normal architecture is critically dependent on mechanisms that ensure that the occasional loss of cells is compensated by the regrowth of a number of replacements. If too little replacement occurs, the tissue will deteriorate and shrivel away. If there is too much, it will expand, push against its normal boundaries, and perhaps erupt into a tumor. Proper control of cell proliferation is critical throughout the life of an organism.

To understand cancer, we must understand how the inner blueprint of normal cells tells them when they should multiply. We must know how that blueprint becomes deranged in the heart of the cancer cell. The roots of cancer lie in that blueprint.

GENES AND MOLECULES: A BRIEF PRIMER

The notion of blueprints implies precision, exactitude, a lack of ambiguity. Chaos can be warded off by carefully drawn blueprints. Long before biologists knew much about the inner machinery of living cells, they realized that such blueprints must exist. At first, blueprints were associated with whole organisms; only later did their importance to the lives of individual cells become apparent.

Gregor Mendel, an Austrian monk, established the principle of organismic heredity in the mid-nineteenth century. He focused on the transmission of genetic traits in pea plants—traits such as flower color and seed shape. His work was forgotten and then rediscovered by three geneticists in the first years of the new century. Mendelian genetics, as it came to be called, rested on several simple concepts. First, all complex organisms, from the pea plant to man, transmitted genes from parent to offspring through identical hereditary mechanisms. Second, the outward appearance of an organism could, in principle, be dissected into a large collection of discrete traits, such as flower color and seed shape in peas, eye color or body height in humans. Third, each of these traits could be traced back to the workings of some invisible information packets that were passed from parent to offspring through sexual reproduction. The efficient transmission of these information packets ensured that offspring developed traits closely resembling those of their parents.

The information packets came to be called genes; each human gene was assigned the role of organizing a distinct body trait. As we learned more and more about genes, it became apparent that all aspects of the human body down to the invisible inner workings of individual cells are dictated by the genes that a person inherits from his or her parents. The master blueprint, it turns out, is no more than a large collection of these genes.

We learned that the blueprinting genes are not stored away in a single central archive located somewhere in the body. Instead, almost every one of the trillions of cells in the body carries a complete copy of the entire blueprint. That simple fact reoriented thinking on how genes organize the construction of complex organisms: Genes act directly to control the behavior of individual cells. Each cell, operating under the control of its genes, collaborates with all the others to create the body's form and function. Hence, the complexity of an entire organism represents nothing more than the aggregate behavior of all of its individual cells. This means that the set of genes governing the lives of cells and the set controlling the shape and behavior of the body are one and the same.

A debate of long standing has swirled around the question of how many distinct information packets—individual genes—make up the human genetic blueprint. The best current estimates range between 70,000 and 100,000. Together, these genes form the genetic library, the master blueprint that is commonly called the human genome.

The fact that the genomic library is segmented into discrete gene compartments has several consequences. As mentioned before, different volumes—distinct genes—can be pulled from its genomic library shelf and read selectively by a cell. In addition, these information packets can become separated from one another as they are passed from a parent organism to its offspring. This helps explain why we inherit only some of the genes carried by each of our parents. The genomic libraries in the fertilized egg become a blend of the genes that both parents carried before.

Still, the depiction of genes as information packets is ultimately unsatisfactory, because this image lacks physical reality. Sooner or later, we must deal with the physical substance of genes. Like all other components of living organisms, genes are concrete objects and therefore must be embodied in discrete, identifiable molecules.

Since 1944, we have known that the physical manifestation of genes is the molecule of DNA. Genetic information is carried in DNA molecules. Their structure is quite simple: Each DNA molecule is a double helix composed of two intertwined strands. Each of these strands is a long polymer formed by stringing together, end-to-end, the single components that, for the sake of this discussion, can be termed bases.

The bases of DNA come in four chemical flavors—A, C, G, and T. Importantly, they can be strung together in any order. The information content of DNA is determined by the sequence of these bases. The stringing together of bases can proceed without limit, resulting in DNA strands that can be tens of millions of bases long. A snapshot of one segment of such a very long strand would reveal a specific sequence of bases such as ACCGGTCAAGTTTCAGAG. Modern gene technology allows us to determine these base sequences, a process called "DNA sequencing." By now, several tens of millions of base sequences have been ascertained for various organisms ranging from bacteria to worms to flies to Homo sapiens.

The total flexibility in the ordering of bases in the DNA means that, in principle, any information, biological or otherwise, can be encoded in DNA molecules. At first glance, an alphabet of only four letters would seem limited in its information-carrying capacities. But in fact four letters are more than sufficient. Morse code, with its three letters (dots, dashes, and spaces), and computer binary code with its two-letter alphabet (0 and 1) also have infinite information-carrying capacity.

The double helix of DNA actually carries two copies of genetic information, one on each of its intertwined strands. Since the time of James Watson and Francis Crick's watershed discovery of 1953, we realize that an A appearing in one strand of the helix always faces a T in the opposite strand; a C in one strand inevitably confronts a G in the other. So the sequence of ACCGGTCAA in one strand will be intertwined with a partner sequence, TGGCCAGTT, in the other.

Since the base sequence of one strand dictates the sequence in the other, the information carried in one strand is present in the other, albeit in complementary language. This redundancy has many advantages. Among the most important is that it enables the helix to be replicated. In particular, the two segments depicted in Figure 1.1 can be separated

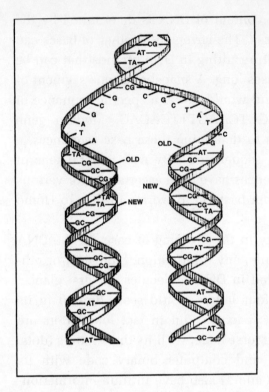

Figure 1.1   DNA replicating itself

and each can be used as a template for copying a new complementary sequence that then becomes wrapped around its template. The result is two daughter double helices that are identical to each other and to the mother double helix from which they arose.

This copying of base sequences becomes important when a cell grows and divides. During the process, a mother cell prepares to endow each of its future daughters with exact replicas of the DNA helices that it carries. This mother-to-daughter transmission allows the genetic information initially present in the DNA of a fertilized egg to be transmitted faithfully through a succession of hundreds of rounds of cell division to all of the trillions of descendant cells that form the adult human body.

Precisely how does the abstract concept of a gene relate to the physical structure of DNA molecules? The double helices of DNA, carried in the chromosomes of the cell, are often hundreds of millions of bases long. These extraordinarily long strings of bases are divided into discrete segments, in effect information compartments, each compartment constituting a single gene. An average human gene takes up several tens of thousands of DNA bases. Certain punctuation marks, written in the four-letter code of base sequences, demarcate the ends of the gene. In an English text, the beginning of a sentence is marked by a space followed by a capitalized word; the beginning of a gene is indicated by a specific short sequence of tens to hundreds of bases. Similarly, like the end of an English sentence, which is set off by a period, the terminus of a gene has its own distinctive base sequence that serves as a clear punctuation mark. Usually, the end of one gene is followed by a sequence of many thousands of bases of meaningless genetic noise before other punctuation sequences signal the beginning of the next gene along the helix.

The total information content of the human genome is carried in about 3 billion bases of DNA sequence divided into the 70,000 to 100,000 segments that represent individual genes. Somehow these genes, working in various combinations inside our cells, are able to create the extraordinary structural complexity of the human body including that organ of infinite complexity, the brain.

This tale of genes, DNA helices, and base sequences provides an entree into understanding all types of human biology, indeed all life forms on earth. But we are concerned here with a narrow slice of this complexity, the disease of human cancer. We can ignore the difficult question of how genes tell cells to collaborate in constructing tissues and organisms, and focus instead on the narrower question of how genes affect the growth behavior of individual cells.

So we narrow our perspective to address the small subset of genes that tell individual cells whether or not they should grow. These genes take us straight to the heart of the cancer problem. They reveal the origins of cancer and will one day point the way to curing the disease.

....

CHAPTER 2

CLUES TO CANCER'S ORIGINS: HOW THE OUTSIDE WORLD AFFECTS THE INSIDES OF OUR CELLS

To understand the roots of cancer, let us leave our discussion of cells and genes and turn to a radically different way of studying and describing human beings and their maladies—the science of epidemiology. Epidemiologists study disease incidence in large groups of people. Cancer epidemiologists, in particular, study the frequencies of cancers in various human populations. Their work is almost always motivated by a central question: How do various kinds of behavior or environment influence the frequency of specific kinds of cancers? The very notion of cancer incidence as an interesting topic for scientific study is relatively new. Until the nineteenth century, cancer was relatively rare, an observation largely explained by the fact that cancer is a disease of older people. In many European countries at the beginning of the nineteenth century, expected life span was only about thirty-five years. Many people who might have contracted cancer late in life were struck down far earlier by infectious disease, malnutrition, or accident.

On the rare occasion when cancer did strike, most attributed it to a random accident or an act of God. But some evidence accumulating after the last decade of the eighteenth century suggested an alternative explanation—that the appearance of a cancer could be tied to specific experiences or lifestyles of the cancer patients. This new mind-set started with physicians who began to document particular kinds of tumors affecting very distinct subpopulations of humanity.

The most famous and perhaps earliest of these discoveries came in 1775 from a London physician, Percival Pott, who described scrotal cancers in men who had worked as chimney sweeps in their youth. Pott's description represented the first instance in which a specific agent or exposure was correlated closely with the appearance of a cancer. Soon after, a surgeon, also in London, reported unusually high rates of nasal cancers in gentlemen who used snuff.

Reports appeared sporadically throughout the nineteenth century reinforcing this theme. Pitchblende miners in eastern Germany were found to succumb to lung cancer, an otherwise rare disease, in large numbers. By the early twentieth century, those working with the newly discovered X-rays were discovered to be at high risk for skin tumors and leukemias. Women who painted luminescent radium on the hands of wristwatches were diagnosed with cancers of the tongue, which was related to their licking paint brush bristles. Beginning in the early 1950s, people who smoked cigarettes were seen to have an increased risk of lung cancer, often twenty to thirty times higher than the nonsmoking population.

Cancer risk was also found to vary dramatically between countries. Liver cancer was eighteen times more frequent in certain parts of Africa than in Great Britain. Stomach cancer struck Japanese eleven times more frequently than Americans. Colon cancer was ten to twenty times more common in the United States than in certain regions of Africa. These dramatic differences were not due to differences in inherited susceptibility. When individuals migrated from one part of the world to another, their children soon assumed the cancer risks typical of their new locations.

The lesson was clear. Spontaneous, unprovoked break-down of the body's tissues no longer provided a persuasive explanation for many kinds of cancers. An alternative theory had become far more compelling: External factors affecting the body—lifestyle, diet, or environment—play an enormous role in determining the onset of disease. This dramatic shift in thinking, which started at the beginning of the twentieth century, coincided with another revolution, this one affect-ing our understanding of infectious diseases. In the last decades of the nineteenth century, Robert Koch and Louis Pasteur found that a variety of lethal diseases could be traced to specific causal agents: bacteria and viruses. Human dis-eases were now portrayed as the consequences of specific, knowable causes rather than as random, capricious acts of nature.

This breakthrough made it possible to redefine and clarify the cancer problem. Now the puzzle of cancer could be stated in more specific terms: How precisely could lifestyle and diet affect the behavior of tissues deep within the body? The solution to this puzzle would have to be described in terms of individual cells, normal and malignant, and within those cells the machinery that drove their growth forward. Such reductionism—distilling complex phenomena down to simple underlying mechanisms—soon became the central theme of modern cancer research and, by the end of the cen-tury, its glory.

CANCER AGENTS AND TARGET GENES

The notion that cancer was not a random, spontaneous de-generation of the body's tissues, but instead was actively in-duced, radically changed the thinking of many cancer researchers. If external agents triggered the disease, perhaps they could be identified and their mechanisms of action studied. Perhaps the entire process from the initial encounter with a cancer-causing agent to the appearance of a cancer could be uncovered. So, toward the end of the nineteenth century, scientists throughout the world tried to recreate cancer in laboratory animals—mice, rats, and rabbits. For years, all such attempts failed.

The first successes came in Japan in the first decade of the new century. Katsusaburo Yamagiwa took instruction from the far earlier studies of European chimney sweeps. Percival Pott's initial observations of high rates of scrotal cancer in London chimney sweeps had been followed several decades later by the work of others who found that chimney sweeps on the Continent had much lower rates of this tumor. The difference seemed to be related to personal hygiene practices. The British chimney sweeps, like many of their eighteenth-century countrymen, rarely took baths, while the chimney sweeps on the Continent bathed frequently. It seemed that creosote tars from the London flues stuck to the skin of the British sweeps and triggered cancer unless washed away quickly.

Responding to this, Yamagiwa took coal tar and applied it repeatedly to the ears of rabbits. After many months, skin tumors appeared on the rabbits' ears. Other researchers had failed to induce cancer because they had given up much too quickly or had not realized the need for repeated applications of the substance.

Yamagiwa's experiment showed directly that cancers could be provoked at will in the laboratory by specific agents. The rabbit ear tumors—and perhaps all cancers—could be traced to well-defined causes. Still, this insight only served to raise another, equally large issue: Precisely how did chemicals, such as those in coal tars, create cancer? Somehow, cancer-causing chemicals—chemical carcinogens—were able to invade cells in the body's tissues and provoke the growth of a tumor. So the metaphor shifted dramatically. Cancer itself was not the invader; rather, the true invader was the carcinogenic agent (in this case coal tar).

The mystery deepened with the observation that X-rays were also capable of inducing cancers. Soon after Wilhelm Roentgen's 1895 discovery, X-ray tubes were used extensively for imaging bones and then for treating a variety of conditions. The technicians who ran the X-ray machines as well as many of the exposed patients contracted skin tumors and leukemias. What hidden connection existed between these two apparently unrelated agents—chemicals and X-rays—that allowed both to trigger cancer? Both were noxious and both were capable of damaging human tissue and killing cells. But how could cell killing be related to cancer? Cancer represented an excess of cells in a tissue, just the opposite from what was seen after toxic agents depleted a tissue of its cells.

By the 1930s, the coal tar question had been focused far more precisely by British chemists working hand in hand with cancer researchers. They found that coal tars were really a mixture of hundreds, perhaps even thousands, of distinct chemicals. So the chemists separated the tars into a number of constituent chemical compounds and handed each over to the cancer researchers, who proceeded to test each compound in laboratory animals for its cancer-inducing ability. Some of these were found to be very potently carcinogenic. Now the mystery of chemical carcinogenesis could be restated in even more precise terms: cancer could be caused by specific chemicals, compounds such as 3-methyl-cholanthrene and dimethylbenzathracene—and, of course, by X-rays.

Still, this incremental step did little to address the fundamental problem of how these or any other chemicals could trigger cancer. As often happens in cancer research, the big leaps forward in solving this particular problem came from work that had no apparent connection with cancer. On this occasion, the most telling ideas came from work on the genetics of fruit flies. By the first decades of the twentieth century, fruit flies were seen to embody a system of heredity very much like that observed in humans.

Even more to the point, the genes of fruit flies were susceptible to alteration. Normally, the offspring of a pair of flies were indistinguishable from the parents. But in the 1930s, Hermann Muller discovered that fruit flies that had been exposed to X-rays would, with some frequency, produce offspring having quite different traits. These novel traits would often be passed on to the next generation of offspring, and beyond that to succeeding generations.

Muller concluded that the genetic material, which had seemed to be transmitted faithfully and accurately from one generation to the next, was vulnerable to change. It was, as the geneticists said, mutable. In some unknown fashion, X-rays could strike at genetic material and change its information content. So the scientific thinking and vocabulary changed: X-rays could mutate genes.

The unpredictable genetic changes wrought by X-rays were usually fatal. But on certain rare occasions, these genetic changes—mutations—did not affect the viability of the flies, which thrived in spite of their altered genes. One well-studied example involved a gene that normally specified a red eye pigment. Following X-irradiation, the mutated gene served as template for an eye that lacked pigment and was, as a consequence, almost pure white. This white-eyed trait could be passed indefinitely from one generation to the next.

By the end of World War II, certain chemicals were also found to induce mutations in fruit flies. Some were highly reactive nitrogen mustards, like those used in the gas warfare during World War I. As before, the immediate offspring and subsequent descendant generations of an exposed fly transmitted an altered version of a gene specifying some discrete trait such as eye color, leg development, or the development of bristles.

Around 1950, several geneticists assembled the body of accumulated information on chemicals, X-rays, and mutations and came up with a grand unifying theory, which in truth was little more than speculation. It went like this. X-rays and certain chemicals can induce cancer. X-rays and cancer can also induce mutations in genes. Therefore, these cancer-causing agents act through their ability to cause mutations in the genes of exposed animals. Stated differently, carcinogens (cancer-causing agents) are really mutagens (mutation-causing agents), and the two processes are inextricably linked.

Implicit in this theorizing was the idea that the genes of flies behave identically to those of humans. By the 1950s, this notion was increasingly attractive. The genes of fly and human cells were both found to be carried by DNA molecules. Also, the cells of all complex organisms, from worms and flies up through humans, were known to be constructed in very similar ways. Extrapolations made from one organism to another thus seemed to rest on very solid ground.

The mutations induced by these mutagenic agents created a bit of a puzzle. Geneticists had studied mutated genes that were passed from one organism to succeeding generations of descendant organisms. In the case of cancer, however, the mutagenic agents seemed to be damaging genes of cells sitting at localized sites throughout the body of a single organism. Once genes in a target cell were damaged, so the thinking went, this mutant cell would begin to proliferate uncontrollably within that organism, sooner or later yielding a horde of descendant cells that was seen as a tumor.

There seemed to be two systems of genetics: one describing the transmission of genes from a parent organism to its offspring, the other describing the transmission of genes from a cell within a tissue to descendant cells within that tissue. In the latter case, the genes that were struck by mutagenic carcinogens usually had no chance of being passed on to the next generation of organisms. Genes carried by cells in the gut or the brain or the lungs, no matter how damaged they were, would never affect the genetic makeup of one's offspring. Only mutational damage inflicted on genes carried in sperm or eggs (in the testes or ovaries) could be transmitted to the next generation.

The dichotomy came to be expressed in simple terms. Mutations in cells of the "germ line" (germ cells being the sperm or eggs) could be passed on to offspring; mutations arising in cells elsewhere in the body (the "soma"), could not. Such somatic mutations, as they were called, were clearly the candidates for the critical events that triggered cancer.

After Watson and Crick's 1953 discovery of the DNA double helix, these speculations about genes and mutations could be recast in more specific terms. If the information carried in a gene was encoded in a sequence of DNA bases, then mutations were nothing more than changes in DNA structure, alterations in the sequence of DNA bases that constituted an individual gene. If the carcinogen-equals-mutagen theory was correct, then cancer cells must carry DNA molecules having altered sequences of bases. These altered DNA sequences, which encoded information not present in a normal cell, somehow directed the cancer cell to grow uncontrollably.

The carcinogen-mutagen theory was attractive because it reduced the complex phenomenon of cancer causation to a simple underlying mechanism. But proving this idea would require another three decades of research. As often happens, genetic theorizing had raced far ahead of the available evidence.

PROVING THAT MUTAGENS ARE CARCINOGENS
....

In the 1930s, it became apparent that a wide variety of chemicals had carcinogenic powers when introduced into laboratory animals. Soon a cottage industry grew up among cancer researchers that focused on inducing tumors in laboratory animals. The animals of choice were usually mice and rats. Like Yamagiwa's rabbits, their biology was reasonably close to that of humans, and they could be maintained in large colonies and exposed over many months to repeated doses of chemicals. This testing of potential carcinogens became ever more necessary as the chemical industry began releasing hundreds, then thousands of new compounds into the marketplace after World War II.

By the 1960s this testing created a large catalog of identified rodent carcinogens. Many were suspected to cause cancer in humans as well, but in most cases this would never be proved, because humans could not be exposed intentionally to suspected carcinogens. Chemicals of demonstrated carcinogenicity in rodents often were pulled from the market, or if they were approved for general use, their applications were severely limited.

These rodent carcinogen tests revealed a great deal about the chemical species that triggered cancer. They demonstrated that a wide variety of chemicals having very divergent molecular structures were potential carcinogens. After entering into the body and its cells, these chemicals would bond with diverse target molecules inside the cells, somehow changing, even damaging them. The diversity of chemical carcinogens suggested a corresponding diversity of target molecules inside human cells.

Another insight came from the observation that chemicals differ greatly in their ability to create cancer in mice and rats. In some cases, hundreds of milligrams of a chemical compound, administered over many months, were needed to trigger cancer. Other chemical species succeeded when introduced into a rat or mouse only once or twice in doses of a few micrograms. These cancer-causing potencies might differ by a factor of a million or more. One of the most potent of the chemicals tested was of natural origin—the compound aflatoxin, which is made by a mold that grows on improperly stored peanuts and grains. Minute amounts of aflatoxin were highly effective in triggering liver cancer in both rodents and, as shown by epidemiology in Africa, in humans.

The bewildering array of chemicals that were cataloged as carcinogens seemed to confound rather than simplify the thinking about cancer's origins. How could this mountain of evidence be boiled down to a small number of simple principles? More to the point, how did the behavior of these chemicals shed light on the carcinogen-mutagen theory?

In the mid-1970s, Bruce Ames, a geneticist working at the University of California at Berkeley, provided one key to this puzzle. Ames had focused his earlier research on the question of how bacterial genes operate. His work, like much of bacterial genetics, had wide impact, because the genes of bacteria function much like the genes of more complex life forms. Bacterial genes are encoded in DNA molecules and, like our own genes, are susceptible to mutational damage. Also, X-rays and many of the chemicals that damage human genes had identical effects in bacteria.

The study of bacterial genes offered one significant advantage over gene research involving humans or mice. Bacteria can be grown in enormous numbers, quickly and cheaply. They reproduce in twenty minutes, unlike mice, which require several months. So gene research in the 1960s and 1970s was driven largely by advances in bacterial genetics.

Ames wanted to develop a simple method of measuring the relative mutagenic potency of various chemicals. The genes of salmonella bacteria, grown in a petri dish, served as the targets of his chemicals. In his most widely used test, the mutation of a critical gene allowed a mutant bacterium to multiply into an easily observed colony in the dish; unmutated bacteria would fail to do so. So the potency of a candidate mutagen could be gauged simply by introducing the chemical into a petri dish seeded with the proper bacteria, allowing the chemical to mutate genes of these bacteria, and counting the number of bacterial colonies that soon appeared in the dish. The number of colonies increased in direct proportion to the mutagenic powers of the tested chemical.

Ames assembled a large collection of known carcinogens and began testing them, one by one, with his bacterial mutation assay. Analysis of his results led to a provocative correlation. Chemicals that were highly potent in inducing mutations in bacteria were also potent in triggering tumors in laboratory rodents; those that lacked substantial mutagenic ability seemed to lack cancer-causing ability.

For the first time, the carcinogen-mutagen theory had some experimental underpinning. It seemed ever more likely that the ability of a chemical to induce cancer derived from its ability to damage genes in the body's cells. There really was a close connection between mutagenicity and carcinogenicity.

The Ames test, as his method came to be called, offered another windfall. Scientists could now gauge the potential carcinogenic powers of newly developed chemical compounds in one or two days. This was cheaper by a hundredfold than the several-year-long rodent tests that had been required until then to test the safety of chemical compounds slated for human exposure. Positive result in an Ames test almost always doomed the future development of such compounds.

Of course, things were not really that simple. Some chemicals that registered negative in the Ames bacterial test turned out to be quite effective in increasing cancer incidence in rodents and humans. Asbestos and alcohol are notable examples. Yet others that were highly effective in mutating bacterial genes were rather weak carcinogens in mammals.

Still, the point was made: Mutagens create cancer through their ability to enter cells and damage genes. Soon a number of carcinogens were shown to react directly with DNA molecules, specifically the bases in the two strands of the double helix. By altering the structure of these bases, they directly affected the information content of the DNA, precisely the behavior expected of a mutagenic agent.

So the carcinogen-mutagen theory gained support, bolstered by the findings of Ames and others that many carcinogens were capable of creating mutant genes through their ability to damage DNA. However, this was only one of many theories on the origins of cancer, and it could never rise above the others to become accepted truth as long as a vital piece of evidence was missing. If carcinogens created cancer by altering genes, then cancer cells must carry mutated genes. These genes needed to be found. Without them, the carcinogen-mutagen theory would eventually be shelved to take its place among dozens of other failed explanations for this complex disease.

....

CHAPTER 3

THE ELUSIVE QUARRY:
HUNTING THE PROTO-ONCOGENE

By the mid-1970s, Ames and countless other chemical carcinogen researchers embraced a simple desciption of the origin of human cancer. Once they took to the new religion, they
began preaching it widely. Their message was that cancer is
caused by chemical and physical agents that damage genes
carried by cells deep inside the body's tissues. Meanwhile,
another school of cancer researchers stood on a different
street corner and preached a diametrically opposite point of
view. They found the chemical research unconvincing. Their
view was that cancer was caused by infectious agents. This
other school portrayed cancer as an infectious disease, spread
by microbes, not by mutagenic chemicals or radiation.

The culprit microbes were viruses, subcellular forms of
life. Virus particles are little more than packets of genes, carried in protein and lipid shells, that travel from one cell to
another. After attaching themselves to the surfaces of cells,
virus particles inject their genes into these cells. The injected
viral genes then begin to copy themselves. Soon these newly
copied viral genes are wrapped up into new packages—new
virus particles, assembled from the chemical building blocks
that the host cell had stored for its own use. Then the viral

progeny particles burst forth from the infected cell and begin searching for new victims to parasitize.

Seen in this way, a virus's sole function seems to be to make more copies of itself. In the course of doing so, a virus and its progeny may kill many cells and damage substantial amounts of tissue. Viruses trigger respiratory infections, rabies, measles, mumps, rubella, smallpox, and cold sores, in each case leaving swaths of destroyed tissue in their wake.

But some viruses show a very different behavior: Rather than destroying tissues, these viruses trigger cancer. Peyton Rous of the Rockefeller Institute in New York discovered the first known tumor virus in 1909. He found that virus particles extracted from a connective tissue tumor in one chicken could induce a tumor when injected into a second bird. Extracts of that tumor, once again, yielded cancer-inducing virus particles. Indeed, Rous's sarcoma virus could be transferred indefinitely from one chicken to the next. The virus would multiply in the cells of each infected bird and trigger the formation of tumors.

In the 1930s, new cancer viruses were found that caused skin cancers in rabbits. Then came a mouse breast cancer virus and yet another virus that induced leukemias in mice. A relative of Rous's sarcoma virus was found to trigger a leukemia-like disease in chickens. By the 1950s, several more leukemia viruses had been found in mice.

The notion emerging from these discoveries was a simple one. A tumor virus would enter a cell somewhere in the body. Instead of multiplying inside this cell and killing it, the virus would allow the cell to live. Stopping short of killing was part of the virus's agenda of establishing long-term residence in its newfound host. Once ensconced within the cell, the virus would begin to tamper with its host's growth-controlling machinery, forcing the cell and its descendants into endless rounds of proliferation. Somehow the virus particles, which were thousands of times smaller than the cells they infected, were able to usurp control of these cells and dominate their behavior. By driving endless cycles of cell growth and division, an infecting tumor virus particle could generate the excess cells that formed a tumor mass.

These takeovers of cells by tumor virus particles presented a clear and powerful explanation of how human cancers are triggered. But there were problems with the attempts to attribute human cancers to tumor virus infections. The most troubling inconsistency came from the epidemiologists, who had shown clearly that most kinds of human cancer do not behave like contagious diseases. When the geographic locales of cancer cases were plotted on maps, they seemed to be distributed randomly across the landscape rather than being localized in small, dense clusters, as might be expected from an infectious disease.

Those who embraced the tumor virus theory could rationalize this inconsistency. They imagined that human cancer viruses were ubiquitous in the human population. Like the bacteria that inhabit the skin and gut, these viruses normally caused little harm. But on rare occasions, provoked in some unknown way, they might erupt and cause great damage by triggering a malignancy. Such widespread distribution and occasional bad behavior were compatible with the observation that cancers appeared as isolated events, striking here and there rather than in large epidemics.

By the early 1970s, those who pushed the candidacy of tumor viruses began to use the new science of molecular biology to strengthen their case. By dissecting cancer viruses molecule by molecule, the tumor virologists began to understand precisely how these viruses succeeded in converting normal cells into cancer cells. Their motive for doing this research was simple: They wanted to make the case of human cancer viruses more plausible.

Like all other organisms capable of making more copies of themselves, tumor virus particles were found to carry a number of distinct genes. Some of the viral genes were dedicated, as expected, to the process of viral replication. These "replicative" genes served as templates that allowed these viruses to make copies of themselves within infected host cells. In addition, cancer viruses seemed to carry extra genetic information that enabled them to transform infected hosts from normally growing cells into aggressively growing cancer cells. Hence, virus-borne genes could trigger cancer.

This finding reinforced the notion that cancer was a condition created by the actions of genes. But it did not help to bridge the chasm separating the chemical and viral theories of cancer. The chemical carcinogenesis researchers argued, as before, that the cancer-inducing genes inside tumor cells were not of viral origin but were indigenous to the cancer cell. They were altered versions of normal cellular genes that had been damaged by chemicals or radiation. Those who believed in viruses were convinced that all tumor cells bore cancer genes of foreign origin—genes that had been forced on these cells by invading tumor viruses.

But one unifying thread did tie the two groups together. Both agreed that a small set of genes operating inside a cancer cell was capable of causing the cell and its descendants to grow without limits. They called such cancer genes "oncogenes," echoing the term *oncology,* which refers to the specialty of treating tumors, and the Greek word *onkos*, meaning a lump or mass.

A GOOD GENE GONE BAD

The debate came to a head in 1976. The warring schools espoused seemingly irreconcilable theories of the origins of human cancer. As it turned out, both lines of thinking contributed vital pieces to the puzzle of cancer's origins. In fact, there was a way to bridge the gap. It went like this. Perhaps oncogenes were indeed important for triggering cancer, but maybe the oncogenes that triggered human cancer were not imported by viruses into cells. Instead, these genes might be native to human cells. Maybe chemical carcinogens acted by damaging normal cellular genes. Once mutated, these cellular genes would be converted into potent oncogenes that operated much like the oncogenes forced into cells by infecting tumor viruses.

This idea was attractive but apparently unprovable. Like most unprovable notions, it was dismissed as speculative and hence scientifically worthless. The halls of cancer research were littered with the corpses of dozens of explanations of cancer's origins. This one seemed destined for a similar fate.

A human cell, normal or malignant, was suspected to carry many tens of thousands of genes in its DNA. Among these might be a small set of genes which, when mutated by chemical carcinogens, would trigger runaway cell growth. For the moment, the task of finding these few mutant genes inside a cancer cell was far beyond the reach of available technology.

Yet the cellular oncogenes were found, ironically by researchers who were studying viral oncogenes. The discovery of these genes triggered the revolution in cancer research that continues to this day.

The key to the puzzle came from Rous's sarcoma virus, often called RSV. After 1909, Rous had abandoned research on his virus, convinced that it held no relevance for understanding the root causes of human cancer. Other researchers studied RSV at irregular intervals over the next sixty years. In 1966, by then in his mid-eighties, Rous received the Nobel Prize in medicine and physiology for the work he had done more than half a century earlier.

Interest in cancer viruses had revived in the 1960s, in part through the recruitment of a new generation of young researchers who were eager to exploit techniques of DNA analysis to dissect these cancer agents. Among these were Harold Varmus and J. Michael Bishop in San Francisco. They wanted to know how RSV grew inside infected chicken cells and, more importantly, how it transformed these cells from a normal growth state to a malignant one.

Varmus and Bishop drew on the work of others who had begun to dissect the small genome of RSV. Like other viruses, RSV carried several genes that it used to replicate itself inside infected cells. These replicative genes instructed the cell to produce hundreds, even thousands, of progeny virus particles identical to the one that initiated the infection.

In the end, however, it was another gene of RSV that stood out. This was the gene used by the virus to transform infected cells from a normal to malignant growth—the virus-borne oncogene. This viral gene was called *src* (pronounced "sark") for "sarcoma." All evidence indicated that after *src* was inserted into a cell by an infecting RSV particle, it would issue marching orders that drove the cell and its direct descendants into endless rounds of growth.

The origin of the *src* gene of RSV presented another major puzzle. Other viruses, close cousins of RSV, shared the same replicative genes and could multiply in infected chicken cells like RSV, but these cousin viruses could not convert infected cells into tumor cells. They also lacked the *src* gene, reinforcing the impression that *src* was the tool by which RSV induced cancerous growth.

Most geneticists, viewing RSV and its cousin viruses, concluded that RSV was the true natural virus and that its cousins were defective mutant versions that somehow had lost the *src* gene and the associated ability to trigger cancer. Viruses were known to lose genes frequently, a consequence of the haphazard operations of their gene-copying machinery.

But the facts in this case seemed to argue otherwise. The cousin viruses were widespread; RSV with its *src* oncogene was unique, having been isolated only once in the early part of the century by Rous himself. It began to seem that RSV was the anomaly and that its cousin viruses represented the norm. Perhaps RSV arose from one of these cousin viruses after it acquired the *src* gene from some foreign source.

Where could the *src* gene have originated? According to the most likely scenario, the ancestor of RSV had lifted the *src* gene from another tumor virus. That acquisition would have allowed RSV to induce cancer, a talent it previously lacked.

The experiments coming from the Varmus-Bishop consortium soon proved genetic theft, but the actual source of the stolen gene turned out to be totally unexpected. Their research group developed a technique for detecting the *src* gene in other genomes, both viral and cellular. With this technique in hand, the researchers began to hunt for other places where the *src* gene might be present.

In 1975, a researcher in their lab came across a most surprising result in the course of a routine experiment. He was using their newly developed technique to analyze the genes of normal chicken cells and those that had been infected by RSV. The expectation here was clear: The normal cells would lack the *src* oncogene, while the infected cells would have at least one copy of this gene, it having been imported into the cell by the infecting RSV.

These expectations were dead wrong. The *src* gene was clearly present in both uninfected and infected cells. Normal chicken cells had at least one copy of *src* long before their infection by RSV!

Their discovery was published the next year, triggering the revolution of 1976. The finding led to a dramatic reorientation in thinking. It suggested that *src* might originally have been a cellular gene that was kidnapped by an ancestor of

RSV, incorporated by RSV into its own genome, and then exploited by the virus to transform normal cells into cancer cells.

The new scenario went like this. RSV had arisen as a totally new virus in the months before Rous discovered it in 1909. Its immediate ancestor was one of the cousin viruses that were able to multiply in chicken cells but unable to transform them into cancer cells. On one occasion, while growing in a cell of an infected chicken, one of these cousin viruses, through some type of genetic accident, had incorporated a copy of the cell's *src* gene into its own viral genome.

The *src* gene appeared to be a normal chicken gene. Prior to this theft, *src* was responsible for some normal aspect of the chicken cell's growth and was not involved in the process of creating cancer. But once it became part of the RSV genome, *src* was subverted by the virus, which refashioned it, causing *src* to undergo a Jekyll-to-Hyde conversion. A normal cellular gene had been reshaped into a potent agent for causing cancer.

Soon it became clear that at least one copy of the normal *src* gene was present in the genomes of all birds. Later the gene was found in the genomes of all vertebrates, including humans. This meant that the *src* gene was part of the normal genetic apparatus of all animals with backbones. Within several years, a version of the normal *src* gene was found in distantly related animals including the fruit fly.

This presence of the normal *src* gene in the genomes of such diverse animals meant that a version of this gene was present in the common ancestor of all these organisms more than 600 million years ago. This gene was retained in the descendant organisms because it played an indispensable role in their lives. If *src* had not been such an important actor, it would have been discarded by at least some animal groups at some point in their evolution. But it appeared to be universal.

So the *src* gene had two incarnations. In its normal guise, it operated in the cells of all animals, templating some essential function. After being acquired by RSV, however, *src* took on the role of an oncogene, enabling RSV to become a potent cancer-causing agent. The association of *src* with RSV represented a rare accident—a genetic theft that happened in a Long Island chicken coop in 1909 several months before the tumor-bearing chicken came to the attention of Rous.

But an even more significant lesson overshadowed these viruses and their genetic versatility. The Varmus-Bishop team called the normal version of the *src* gene a "proto-onco-gene," indicating that it had the potential, revealed under the proper circumstances, of converting itself into a potent cancer gene, an oncogene. Their term implied that there was at least one latent cancer-causing gene hiding out in the genomes of chickens and, by extension, in the human genome as well.

Thinking about the origins of cancer underwent a revolutionary change. For the first time, it became plausible that the roots of the disease lay deep within normal cells. Each cell seemed to carry within its normal genome the seed of its own destruction in a gene that it used to carry out its normal, everyday business.

THE ENEMIES WITHIN

Soon the Varmus-Bishop laboratory and others began examining other viruses that, like RSV, had the ability to create tumors. These other viruses infected chickens, mice, rats, monkeys, even cats. They were all distantly related to one another, being members of the retrovirus class. (Years later, when HIV, the agent of AIDS, was discovered, it would be found to be a distant relative of these cancer-causing retro-viruses.)

The various cancer-causing animal retroviruses all turned out to have histories that were strikingly similar to that of RSV. Each had arisen from an ancestral virus that lacked the ability to induce tumors rapidly in infected host animals; each had acquired potent cancer-causing ability by picking up a proto-oncogene from an infected host—whether chicken, mouse, rat, monkey, or cat. The newly acquired genes were remodeled by these viruses into potent onco-genes.

The proto-oncogenes stolen by these various viruses were different from *src*. Each received its own genetic name— *myc, myb, ras, fes, fms, fos, jun.* These names reflected the viruses in which each of these genes was first discovered: *myc* was first isolated in avian myelocytomatosis virus, *ras* from a rat sarcoma virus, *fes* from feline sarcoma virus. The list soon grew to more than twenty genes.

Now the example provided by *src* could be extended and generalized: The genomes of animals contained many proto-oncogenes, most unrelated to *src*. Each of these, like *src,* had widespread distribution through the animal kingdom. There-fore, *myc* and *myb,* originally found in chicken DNA, were represented in the DNA of all mammals. It seemed that the entire catalogue of proto-oncogenes was present in the genomes of all backboned animals.

This meant that the human genome carried a large set of latent cancer genes—proto-oncogenes that played an essen-tial role in the life of human cells. That much was made clear by the conservation of these genes, in almost unaltered form, in the genomes of animals over hundreds of millions of years of evolution.

It would require the better part of a decade to uncover the precise functions of these normal genes. For the moment, their normal roles seemed a bit of a distraction. The discovery of these genes had a larger, more obvious impact: They represented candidates for the targets of chemical carcinogens. Proto-oncogenes in animals were occasionally activated by retroviruses into oncogenes; maybe the same genes in humans were also activated by mutagenic carcinogens. Instead of being stolen away from their normal roosting sites in the cell's chromosomes and remodeled by passing retroviruses, these genes might be changed on site by attacking carcinogens. The end result could possibly be the same—the creation of powerful oncogenes.

So, within several years in the mid-1970s, a seemingly impossible problem had the prospect of being solved. The retroviruses had made this all possible because of their constant, promiscuous scavenging of host cell genes which occasionally netted them big fish—proto-oncogenes. These viruses had opened a window on these genes, only several dozen among the many thousands of genes in the human genome. The proposal that proto-oncogenes played critical parts in triggering human cancer still needed validation, but even without that hard evidence, it encouraged those who wanted to believe that the roots of cancer could indeed be found in our genes. Soon they came across an abundance of evidence that exceeded anyone's expectations. That evidence converted an attractive hypothesis into a rock solid truth.

••

CHAPTER 4

FATAL FLAWS:
FINDING ONCOGENES
IN HUMAN TUMORS

The 1976 announcement of the discovery of the *src* proto-oncogene suggested an obvious next step to researchers. If this gene were indeed present in the normal human genome, then human tumors might carry *src* in a mutant, actively oncogenic form. To be sure, the mechanism creating such an activated *src* oncogene might be very different than that used by Rous sarcoma virus. By 1976, half a dozen years of fruitless searching had convinced most researchers that retroviruses like RSV were absent from virtually all human tumors. Hence, so the thinking went, if *src* had been activated into an oncogene in these tumors, nonviral agents such as chemicals must be responsible for its activation. These chemicals would mutate *src* while it sat in its normal position in one of the cell's chromosomes. Each time this occurred in a human cell, the resulting *src* oncogene would trigger runaway growth, allowing the mutant cell to generate the large host of descendants that eventually appeared as a tumor. So the researchers surveyed human tumor DNAs for mutant *src* oncogenes. They came up empty-handed. The mutant forms of *src* were nowhere to be found. The *src* oncogene began to look like a quirk of RSV. For the moment, the simple notion that proto-oncogenes were targets of carcinogens seemed to be foundering.

By 1979, another strategy for searching for elusive tumor oncogenes came online. This new approach did not depend on the knowledge gained about retroviruses. It was an independent strategy made possible by the experimental technique of gene transfer. Simply put, gene transfer made it possible to extract DNA (and thus genes) from one cell and introduce these genes into a second cell. The genes introduced into the recipient cell might cause it to take on new traits or behaviors. Such a response would indicate that the information specifying the newly displayed trait was present in the donor cell (from which the DNA had been prepared) and that this information could be conveyed to a recipient cell by the transfer of DNA molecules.

Gene transfer was used to search for oncogenes that might be present in the genomes of mouse, rat, and human tumor cells. None of the tumors from which these cells derived had any viral associations. Most were known or suspected to have been caused by carcinogenic chemicals.

In an early experiment, undertaken in my laboratory, DNA was prepared from mouse cells that had been transformed into cancer cells by exposure to a coal tar carcinogen. This DNA was then introduced into normal mouse cells. The hope was that information dictating malignant growth might be present in the DNA of the chemically transformed cells. When this information, in the form of specific genes, was transferred into the normal cells, the latter might respond by transforming themselves into cancer cells.

We soon found that some of the recipient cells that had taken up tumor cell DNA became transformed! That conversion of normal cells into malignant cells proved that the information causing these cells to grow malignantly was carried in DNA originating in the chemically transformed cells. Information for cancerous growth could indeed be transferred from one cell to another via DNA molecules.

Soon, human tumor DNAs were also discovered to harbor such cancer genes. Research in Geoffrey Cooper's and Michael Wigler's laboratories, together with continued work by my research group, extended the list of tumor genomes carrying information for malignant growth: A bladder carcinoma, a colon carcinoma, and then a nerve cell tumor yielded DNA having transforming activity. In each case, the introduction of tumor cell DNA into normal mouse cells caused these cells to transform themselves into cancer cells. However, DNA from normal cells lacked this transforming activity.

This transferable genetic information behaved much like the oncogenes carried by cancer viruses. In particular, discrete segments of DNA appeared capable of redirecting the metabolism of normal cells, forcing them to become cancerous. These discoveries provided direct indications that normal cells carried genes that were capable, after being mutated, of causing cancer.

The mutational processes that created these oncogenes seemed to mirror the processes that damaged many other kinds of cellular genes: The sequence of bases was changed while the genes sat in their normal positions in the chromosomes. Moreover, following mutation, these genes remained in their normal positions but began to issue instructions that were quite different from their earlier messages to the cell.

This scenario differed dramatically from the story inspired by the retrovirus research. All the transforming retroviruses seem to have originated from progenitors that swooped into normal cells, kidnapped normal cell genes, and converted them into active oncogenes. In the case of these retroviruses, the activation of oncogenes seemed to depend on the subjugation of normal cellular genes by foreign invaders, powerful viral regulators that subverted these genes and then carried them far away from their normal chromosomal roosts.

Still, there was a common thread: Both the cell- and virus-associated oncogenes arose from normal cellular precursor genes—proto-oncogenes. This similarity provoked an obvious question: What was the relationship, if any, between the proto-oncogenes mutated on site by chemicals or radiation and the proto-oncogenes kidnapped and activated by retroviruses?

The answer came in 1982. By then, some of the human tumor oncogenes had been isolated by the newly developed procedure of gene cloning. An oncogene isolated from a human bladder carcinoma was compared with the large cohort of proto-oncogenes that had been cloned earlier by the retrovirus researchers. The comparisons yielded a striking connection: The human bladder carcinoma gene was virtually identical to the *ras* oncogene that retrovirus researchers had found in a rat sarcoma virus.

Suddenly, many different pieces in the puzzle of cancer fell together. The story went like this. While infecting rat cells, a retrovirus had acquired and activated the *ras* proto-oncogene, much as the *src* gene had been acquired by the precursor of RSV. This activated *ras* oncogene, borne by the agent that came to be called Harvey sarcoma virus, was able to transform normal rodent cells into aggressively growing tumor cells.

It was no surprise that a relative of the *ras* proto-oncogene was also present in almost identical form in normal human DNA. By the early 1980s, it had become clear that all proto-oncogenes were uniformly present in the genomes of all mammals and birds.

The *ras* proto-oncogene in a human bladder cell had encountered a different fate than its relative in the rat genome. Some mutagenic chemical had entered the bladder cell and mutated this *ras* proto-oncogene, converting it into an activated oncogene. Once mutated, this *ras* oncogene drove the proliferation of the mutated cell and all of its lineal descendants. The result was a large population of bladder carcinoma cells bearing the mutated, oncogenic *ras* gene.

So the repertoire of proto-oncogenes discovered by the retrovirologists did have direct relevance to human cancer. The same normal genes that were activated in animals by retrovirus capture and remodeling could serve as targets for mutagenic chemicals in humans. While sitting in place amid the chromosomes of a target cell, these human genes could be altered by mutagenic molecules, remaking them into potent oncogenes.

The *ras* gene was only one among many. Within several months, altered forms of the *myc* proto-oncogene were found in human lymphomas and leukemias; this gene was first known from its liaison with the chicken myelocytomatosis virus. Later, a close relative of *myc,* called N-*myc,* was found in neuroblastomas; then *erb* B, first discovered through its association with a chicken erythroleukemia virus, was found in altered form in human stomach, breast, ovarian, and brain tumors.

The plot seemed to be getting much simpler. All vertebrate cells seemed to carry a common set of proto-oncogenes. These genes could become converted into potent cancer-causing genes either by retroviruses or by nonviral mutagens. Proto-oncogenes seemed to represent the ultimate root causes of cancer.

MUTATIONS
. . . .

The discovery of human tumor oncogenes and antecedent normal genes had unified previously unconnected research, including the extensive work on animal retroviruses, but a gaping hole still remained in the evidence. Precisely how did nonviral mutational processes convert normal proto-oncogenes into virulent oncogenes?

The first answers came in late 1982 when researchers compared the human bladder carcinoma oncogene with its precursor, the normal human *ras* gene. It was immediately clear that the search for the mutation that distinguished these genes would not be quick and easy. The two genes were outwardly very similar. They were both five thousand DNA bases long and had identical sequence punctuation marks scattered along their length. This ruled out one possible explanation of their differences—that the conversion from proto-oncogene to oncogene came about through some large-scale deletion or rearrangement of DNA sequence.

But the two genes had to differ in some significant way. When inserted into normal cells, the proto-oncogene had no obvious effect, while the oncogenic version rapidly forced these cells into malignant growth. A detailed, base-by-base analysis of DNA sequences was called for, because the differences clearly were going to be extremely subtle.

The answer, when it finally came, was stunning. The two gene versions, each five thousand DNA bases long, were identical except at one base! In one region, the sequence of the normal gene read GCC GGC GGT, while the corresponding sequence of bases in the oncogene read GCC GTC GGT. A single G present in the normal version of the gene had been replaced by a T in the gene carried by the bladder carcinoma cells. That tiny change in sequence—termed a point mutation—sufficed to change the meaning of the entire gene. It was as if a whole book chapter had its meaning completely changed by the word *dear* being accidentally misprinted as *dead*.

Now the sequence of events that led to the appearance of the bladder carcinoma could be pieced together. The tumor had arisen in a fifty-five-year-old man who had smoked for thirty years. Like all other smokers, he had filled his lungs with potently mutagenic carcinogens. Some of these had been detoxified in his liver, while others had been passed through the kidneys into the urine. Some of these potent carcinogens proceeded to attack the cells lining the bladder, entering them and striking randomly at their DNA. In one of these cells, a *ras* proto-oncogene had been damaged through the change of one of its bases from a G to a T. The altered *ras* gene, now an active oncogene, began to drive the growth of this cell. Years, perhaps decades, later its descendants, all carrying this mutated *ras* oncogene, appeared as a life-threatening tumor mass.

Soon the mutational mechanisms behind other human tumor oncogenes came to light. Each oncogene had undergone its own type of mutational alteration. The point mutation found in the *ras* oncogene turned out to be the most subtle of the changes. In some human tumors, the *myc* oncogene or its cousin, N-*myc,* were found to be present in multiple copies, sometimes as many as ten or twenty per cell rather than the two copies found in normal cells. As a result of these "gene amplifications," tumor cells appeared to experience fluxes of growth-stimulating signals proportional to the number of excess gene copies present.

In certain tumors arising from the lymphocytes of the immune system, notably Burkitt's lymphomas, the *myc* proto-oncogene experienced a quite distinct mutational change. Through a process of chromosomal breakage and rejoining, DNA segments that were previously not associated with one another became fused. As a result, a normal *myc* gene sitting on one chromosomal arm became fused to genes responsible for making antibody molecules. The unnatural alliance subverted the *myc* gene, which was now forced to operate under the control of the antibody gene. The readout of the *myc* gene, previously finely modulated, was now driven at unrelenting high levels, turning it into a potent oncogene.

The overall lesson was clear: Each proto-oncogene became converted to an oncogene through its own distinct mutational mechanism. The identities of the agents or forces that provoked these mutations were still elusive but would be uncovered in the years to come. Independent of cause, the consequences for cells were clear. Once cells acquired activated oncogenes, their normal growth program would be derailed by the strong growth-promoting signals issued by these genes. Major pieces of the cancer puzzle had fallen into place.

CHAPTER 5

A BOOK OF MANY CHAPTERS: MULTISTEP TUMOR DEVELOPMENT

The story of how cancer begins seemed so very simple. A mutagenic chemical invaded a cell, struck a critical proto-oncogene, and converted it into an oncogene. The cell, responding to the orders issued by the oncogene, began a program of unconstrained proliferation. Copies of the oncogene would be passed on to all of the descendants of the initially mutated cell, driving them into forced marches of unrelenting growth and division. Eventually, after some years, a throng of billions of cells would accumulate to form a life-threatening tumor. Those who liked to reduce complex processes to simple explanations were very pleased by this idea. Here was a clear example of how research in molecular biology could uncover a hidden mechanism of great simplicity—so simple and logical that scientists would describe it as "beautiful."

But some thought the scheme was much too simple. They even called it simplistic, and implied that those who believed in it were intentionally ignoring much that was known about cancer formation. In the eyes of these skeptics, the 1982 discovery of the bladder carcinoma point mutation, made in the author's laboratory and those of Mariano Barbacid and Michael Wigler, had caused a rush to unwarranted conclusions.

The skeptics cited a mountain of evidence showing that cancer formation is a complex process involving a long sequence of steps rather than a simple one-hit event that converts a fully normal cell into a highly malignant one in a single step. The remaining years of the 1980s were spent reconciling these two conflicting points of view, the one portraying cancer formation as a simple conversion of the normal cell to a malignant derivative in one fell swoop, the other depicting it as a complex process of many events.

Some of the most persuasive evidence for greater complexity came from the epidemiologists who measured the rate of cancers in various populations at different ages. Colon cancer appeared to occur as much as a thousand times more frequently in seventy-year-olds than in ten-year-olds. The incidence of most other adult cancers also increased steeply with age.

Right away, a simple scenario depicting a single-hit process of cancer formation was rendered much less likely. If cancer was triggered by only a single event, that event would occur with comparable likelihood throughout life. The chance of its happening on any day during the tenth year of life would be no less than on any day during the seventieth. The mathematics of such equal risk led to clear predictions: The risk of a person's having contracted cancer at some point in his or her life, plotted as a function of age, would be a rising straight line. A twenty-year-old would have twice the accumulated risk of a ten-year-old, and a seventy-year-old would have seven times the risk.

Such straight lines bore no resemblance to the steeply rising age-dependent curves the epidemiologists were reporting. Early in life, the lines in their graphs describing cancer risk were relatively horizontal. But then, as they measured tumor incidence in populations of progressively

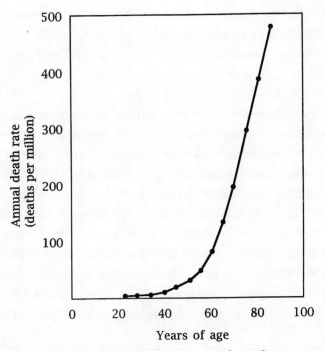

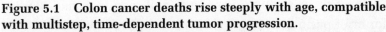

Figure 5.1 Colon cancer deaths rise steeply with age, compatible with multistep, time-dependent tumor progression.
(From U.S. Department of Health, Education, and Welfare. *Vital Statistics of the United States*. Volume II. *Mortality*. U.S. Government Printing Office, 1968.)

greater age, the line turned sharply upward with an ever-steeper slope.

Sharply rising curves like these describe processes of great complexity. They indicate that a multitude of events, happening in succession one after another, must occur before a result is achieved, in this case a diagnosed tumor. The appearance of most cancers seemed to require four to six such events. Each event was itself an occurrence of low probability that seemed to require many years to take place. Only when all of these steps were completed would the process culminate in a clinically detectable malignancy.

The chance that all of these events (whatever they were) would happen in the short period of a young person's life was astronomically small, explaining the virtual absence of most kinds of cancer in children. But as the human body aged, the chances would increase rapidly that all of the necessary events—random accidents—would converge on a cell somewhere in the body. Only then would all the requirements for making a cancer be fulfilled.

Such a drawn-out process of cancer development helped to explain many well-established observations about adult cancer. The best known of these concerned lung cancer. The disease was almost unknown in women at the beginning of the twentieth century and remained rare in women until mid-century. After World War II, women in the United States began to smoke in large numbers, many having started the habit while working in factories during the war. A quarter of a century later, those women began to succumb in large numbers to lung cancer. The process needed to create lung tumors from beginning to end took decades to complete.

Even more striking was the fate of men who had worked for brief periods in naval shipyards during World War II and in the decade thereafter. Many of them were exposed to high levels of asbestos, which was used to insulate components in the hulls of naval vessels. After a time lag of two, three, even four decades, these men, most of whom were cigarette smokers, now began to die from a rare cancer that struck the outside lining of their lungs—mesothelioma. Almost invariably, this type of cancer could be tied directly to asbestos exposure. As was the case with lung cancer and smoking, decades were required between the initial exposure and the appearance of a life-threatening growth.

Epidemiological observations like these made the idea of many protracted steps in cancer formation very persuasive. This notion was attractive because it suggested that the normal human body erects many obstacles to prevent the development of a cancer. Only when all of these obstacles were surmounted, one by one, would a tumor appear.

Still, this multistep theory of cancer formation was hard to reconcile with the 1982 discovery, which suggested that a single dramatic incident such as the mutation of a *ras* oncogene could create a highly malignant cell that proliferated directly into a full-blown tumor. Those who embraced cancer epidemiology dismissed this single-hit oncogene theory as naive and simpleminded.

So two directly conflicting theories of cancer formation arose. The believers in single-hit oncogene mutations took solace from the notion that the science of cancer epidemiology addressed issues far removed from the molecules inside cells and tissues. Did normal cells really need to undergo multiple changes before they became malignant? Or was the epidemiological argument just another dry mathematical abstraction that had little connection with the real biology of human cells?

PLAYING WITH CELLS

Researchers working with genes and cells were amused by the epidemiology but hardly convinced. Hadn't they succeeded in converting normal cells into cancer cells by introducing single oncogenes into them? A single oncogene arose in one hit through the actions of a mutagenic carcinogen. Therefore, inescapably, a single step was enough to create the malignant cancer cell.

But there was a fly in the experimental ointment. Some laboratory workers went back and reexamined the experiment in which oncogenes introduced by gene transfer could convert normal recipient cells into cancer cells in a single, simple event. As it turned out, there was a detail in these experiments that most researchers seemed to have overlooked.

This detail concerned the cells—in this case, mouse connective tissue cells—that had been used as recipients in the gene transfer experiments designed to detect oncogenes in human tumor DNAs. Without doubt, these mouse cells could be transformed into tumor cells in a single step, but were these cells truly normal at the moment these gene transfer experiments began? Or had they already progressed part of the way down the road to being cancerous?

In fact, the skeptics had a point. The mouse cells used in the gene transfer experiments were a bit out of the ordinary. Years earlier, these cells had been taken from mouse embryos and adapted to grow in petri dishes in the laboratory. This adaptation allowed them to be propagated indefinitely. When a population of these mouse cells filled up the bottom of a petri dish, some cells would be transferred to another empty dish and their cycle of growth would begin anew. This process could be repeated indefinitely. The cells used to screen tumor DNAs for the presence of oncogenes had passed from dish to dish for more than a decade before the human bladder carcinoma oncogene was inserted into them.

Cell biologists spoke of cells that could be propagated indefinitely in the lab as being "immortalized." In saying this, they implied that cell immortality was not the normal state of affairs. Most fully normal cells, if taken out of a mouse embryo and placed in a dish, will double only for a limited number of cycles before they stop growing. Populations of mouse cells usually stop expanding after thirty or forty generational doublings in the culture dish; such cell populations are, for that reason, considered "mortal."

On rare occasion, a population of mortal cells will sprout a variant subgroup of cells that have gained the ability to grow without limit. These cells have become immortalized. Curiously, virtually all types of cancer cells also seem to be immortalized. When they are extracted from a tumor and planted in a petri dish, they double without limit. This observation suggested that immortalization occurred routinely during tumor development, perhaps as an essential part of this process.

This rang an alarm among the scientists reexamining the gene transfer experiments. They concluded that the initial experiments, which had been designed to gauge the cancer-inducing abilities of oncogenes, were flawed from the outset because they had been conducted with already-immortalized mouse cells. These mouse cells had undergone an apparent premalignant change long before oncogenes were transferred into them. Maybe they were already poised on the edge of a cliff, and the introduced oncogenes only pushed them over the edge into malignancy.

This idea was given its acid test in 1983. Oncogenes were inserted into truly normal cells—rat cells that had been growing inside a rat embryo only days before. These cells had not been given the chance to develop abnormally through an extended sojourn in petri dish cultures. They were as close to normal as possible, and of course they were mortal.

The skeptics were right. These fully normal cells could not be transformed into cancer cells by single introduced oncogenes, even the potent *ras* bladder carcinoma oncogene. This meant that mortal cells, unlike their immortalized counterparts, were unresponsive to the introduced oncogene. Something needed to happen to them before they could be pushed over the cliff by an oncogene. They needed to be primed to undergo malignant conversion, perhaps by being immortalized. Only then would they respond to an oncogene by becoming cancer cells.

This result suggested that the transformation of a fully normal cell into a bona fide cancer cell entailed at least two distinct changes: conversion of the normal cell into one that was immortalized, and then conversion of the immortalized cell into a malignant one. So cancer involved a succession of at least two changes in the cell, maybe far more.

Only later was it found that the first of these steps—immortalization—could be mimicked or at least promoted by certain other oncogenes such as the *myc* or E1A oncogene. This connection led to another idea: Perhaps introducing two distinct oncogenes into a normal cell might achieve the full conversion to malignancy. Each of these oncogenes might contribute one of the two changes required to make a cancer cell.

So experimenters in my laboratory and that of Earl Ruley began inserting pairs of oncogenes simultaneously into rat embryo cells. Only then did they begin to observe malignant conversion of rat embryo cells. When a DNA clone carrying the *myc* oncogene was introduced simultaneously with one carrying the *ras* oncogene into fully normal rat embryo cells, they responded by transforming themselves into cancer cells! Neither of the two oncogenes introduced singly had this effect.

Thinking about the genesis of cancer cells could now be crystallized around the properties of certain oncogenes. The *myc* and the *ras* oncogenes, each insufficient by itself, were able to collaborate to create cancer. Such collaboration implied that each oncogene acted in its own distinct way to change the cell.

The multistep origin of cancer now became more concrete. Perhaps each of the steps involved in creating a cancer cell represented a rare mutation that affected one or another proto-oncogene in the genome of a cell. Only when two or more such mutations had accumulated would the cell's growth become fully decontrolled.

The model of collaboration between *myc* and *ras* was soon extended to other oncogene pairs. Together, these successes left the impression that two gene mutations might suffice to generate most kinds of cancer cells. But even this number turned out to be an illusion, a vain hope that things would turn out simply. As the 1980s progressed, it became clear that human tumor cells carried many more than two mutant genes, perhaps as many as half a dozen. This higher number, obtained by detailed molecular analyses of tumor cell genomes, seemed to correspond more closely to the number of steps the epidemiologists had inferred from the steepening curves of cancer incidence in aging human populations.

Now the theory of cancer formation could be reformulated: The sequence of rare events that led to a human tumor involved a succession of mutations that progressively altered the genetic makeup of a cell, pushing it further and further toward runaway growth.

CHAPTER 6

FUELING THE FIRE: CARCINOGENS THAT AREN'T MUTAGENS

The notion that human tumor development depended on a succession of gene mutations was enormously satisfying because it echoed a theme that had been reverberating in the halls of science for more than a century. Tumor development showed striking parallels to the evolution of species. In the mid-nineteenth century, Charles Darwin had described evolution in terms of nature's ability to select the fittest from among a heterogeneous population of organisms. After the discovery of gene mutations in the 1920s and 1930s, Darwin's theory of natural selection was refined and extended. Now scientists realized that randomly occurring mutations created genetically heterogeneous populations of organisms, and that natural selection chose among these, favoring the survival and reproduction of those organisms that happened to carry the most favorable constellations of genes.

An analogous process seemed to be playing itself out within human tissues. In this instance, the competing life forms were individual cells. A cell that happened to sustain a mutation altering one of its growth-regulating genes might have a growth advantage over its genetically normal neighbors. It would spawn a host of descendants which would accumulate in disproportionate numbers in the tissue. Later, another mutation occurring in one of these descendants would generate a cell having even greater growth potential, allowing this cell to generate a more aggressively growing flock. These cells would be even more effective in elbowing out their neighbors, outcompeting them for the limited space and nutrients within a tissue.

This evolution within a living body departed from natural Darwinian evolution in one important respect: The continual genetic improvement of the evolving population would eventually compromise its own long-term viability by destroying the environment that nurtured it. Sooner or later, evolving cancer cell populations would kill the host organism that was vital to their own survival.

Still, a vital element in this picture was missing. Each of the mutations implicated in creating a human cancer, such as those generating active oncogenes, was a very improbable event. The chance that a mutation would strike a growth-regulating gene and convert it into a version conferring advantage on the evolving cancer cell was very small—less than one in a million per cell division. Moreover, the number of these mutations required to make a tumor seemed to be quite large—half a dozen and maybe even more.

Following each critical mutations, the descendants of the recently mutated cell would need to multiply into a flock of a million or more before the next one-in-a-million mutation became likely in one of its descendants. This expansion in cell population might take several years to a decade, explaining the long intervals between the successive steps in the process of tumor formation.

The apparently long interval between these steps meant that the entire multistep process was unlikely to reach completion in an average human life span. Yet humans contracted cancer in substantial numbers. Between a fifth and a quarter of all deaths in the West were connected in some way with malignancies.

The attempts to resolve this paradox led to an interesting speculation: Maybe the assumptions about the rates of forward progress were incorrect. More to the point, maybe there were conditions that could accelerate the rate at which these successive steps in tumor formation occurred.

Speculations like these provoked a close examination of the rates with which mutations struck and, deeper down, the molecular mechanisms that created cancerous mutations. At one level, it was obvious that agents such as X-rays and mutagenic chemicals could strike at the DNA double helix and damage its bases, perhaps, as mentioned earlier, by causing one base to be replaced by another or a whole segment of DNA to be deleted altogether.

The rarity of cancer-causing mutations stemmed from the inefficiency of mutagenesis. The responsible agents—chemical and physical mutagens—attacked a cell's genome randomly. Since the important target genes such as the proto-oncogenes represented only a minute fraction of the genome, the mutagens would find these crucial targets only rarely. These jackpot hits would have enormous consequences for the cell, but the likelihood of their happening in any defined period of time was vanishingly small.

Importantly, mutations appeared to happen at a low but constant rate even without exposure to mutagenic agents. These mutations were seemingly spontaneous and found to be intrinsic to all life forms. Indeed, the evolution of species has depended on the slow, spontaneous change in the base sequences of their DNA. Such mutations, occurring continually since life first appeared on this planet, have generated genetic variability and varied characteristics within species. Natural selection has then favored the survival of the genetically best-endowed members of each species. Agents such as mutagenic chemicals and radiation only serve to accelerate the rate at which mutations happen, making them much more likely to occur in any window of time.

A heavy smoker of cigarettes, for example, might collapse the time usually needed to mutate a gene from a decade to a year simply by flooding his or her cells with potent mutagens. Consequently, the entire process leading to a lung or bladder cancer, which might take hundreds of years to complete in a nonsmoker, could be compressed to a few dozen years in this smoker.

But once the actual agents favoring human cancer were identified, it became apparent that this scheme required more refinement. Some chemical agents accelerated cancer formation but did not seem to attack DNA. They were, in other words, poor mutagens. For example, alcohol, asbestos fibers, and estrogen were all known to increase the risk of certain kinds of cancer, sometimes substantially, yet none of these seemed capable of damaging DNA. How could nonmutagenic agents such as these accelerate cancer formation?

The answers came from reexamining the ways DNA is copied inside living cells. When the DNA double helix was first revealed by Watson and Crick in 1953, its structure seemed perfect and robust, well-designed to resist most of the disruptive influences that might be present inside living cells. For example, the bases in the double helix are turned inward and thus are not very susceptible to direct attack by chemical mutagens. Moreover, the linkages between adjacent bases were found to be resistant to cleavage by alkaline ions that arise continually in the cell.

But while the double helix itself is relatively resistant to chemical attack, the process of maintaining a cell's genetic integrity has a weak link. The vulnerability derives from the need to replicate the cell's genome each time the cell goes through the process of growth and division. The resulting duplicate copies of the genome enable the mother cell to en-

dow each of its daughters with a genome precisely equiva-
lent to the one that it carries itself.

This process of DNA replication has flaws. On occasion, a
cell will miscopy a sequence of its DNA prior to cell division, and as a consequence, one of its daughters will receive
a slightly miscopied genome, in effect a mutated one. Even
the best-functioning cells will occasionally miscopy one in a
million (or ten million) bases during each cycle of DNA
replication. Hence, cell growth and division create vulnerability to mutation.

This imperfection suggested another way cancer formation might be accelerated. Agents that promote cell growth
will indirectly create mutations simply because they force
cells to replicate their DNA. More DNA copying means more
inadvertent copying mistakes, hence more mutations.

Knowing this, we began to speculate how some kinds of
carcinogenic agents could work even without having direct
DNA-damaging abilities. An oft-cited example was that of alcohol, which itself seemed to lack mutagenic powers.
Nonetheless, alcohol was a strong carcinogen when con-
sumed in conjunction with tobacco. Repeated exposures of
high concentrations of alcohol were known to kill many of
the cells lining the mouth and throat. The surviving cells in
the tissues lining these cavities would then receive orders to
grow and divide to replace their fallen comrades. These repeated rounds of growth and division would yield mutations
in the DNA of these cells. Moreover, it seemed that DNA in
the midst of replication was even more susceptible to damage from mutagens than DNA from nonproliferating cells.
This explained why cigarette smoke, which contains dozens
of different mutagens, and alcohol, which promotes cell proliferation, were a deadly combination. When used together,
they generated as much as a thirtyfold increase in risk of
mouth and throat cancer.

Study of liver cancer, a leading cause of death in Asia, revealed a similar mechanism at work. Epidemiological studies indicated that its occurrence is closely tied with chronic, often lifelong infection with hepatitis B virus (HBV). In one analysis of government bureaucrats in Taiwan, those having chronic HBV infections were found to run risks of liver cancer as much as one hundred times higher than their uninfected colleagues.

Unlike RSV, HBV does not carry an oncogene in its DNA, and it seems to have little if any direct mutagenic effect on the cells it infects. But it does cause constant and widespread cell killing in the livers of infected individuals. HBV-infected individuals survive for many decades because of the continual replacement of dead liver cells, achieved through the growth and division of uninfected survivors. This constant proliferation provides a stark contrast with the liver of uninfected individuals where cell division occurs only rarely. Once again, an agent can favor the appearance of cancer simply by promoting repeated rounds of cell division.

Estrogen is a fully natural hormone, native to the body, yet it contributes to carcinogenesis in the breast and ovaries. In the breast, it drives the proliferation of cells lining the milk ducts during the menstrual cycle and pregnancy. The monthly multiplication of these mammary epithelial cells is followed by their die-off, this cycle repeating itself over and over in most women from menarche to menopause—generally between ages twelve and fifty.

Many researchers trace the roots of breast cancer to these repeated bouts of estrogen-driven proliferation. The increased incidence of this disease in modern times seems connected to dramatically increased menstrual cycling. Due to greatly improved nutrition, menarche begins four or five years earlier in late-twentieth-century girls than it did in their great-grandmothers. In addition, reproductive practices have changed in Western society. Childbearing and breast feeding, both of which suppress menstrual cycling, are now postponed and, when they occur, encompass only a few years of adult life, unlike a century ago, when three decades of a woman's life were often involved in cycles of birth and lactation. As a consequence, the breast cells of a modern eighteen-year-old may have experienced as many estrogen-driven proliferative cycles as her great-grandmother's breast tissue experienced in an entire lifetime. Once again, conditions favoring cell proliferation contribute importantly to the appearance of tumors. (Independent of these effects are the still-unexplained protective effects of early childbearing and lactation, which reduce lifelong risk of breast cancer.)

These various stories converged on a common theme. By driving cells to grow, some agents could force the process of cancer forward. Cells that grew and replicated DNA were bound to make mistakes in the DNA copying process. More mistakes meant more mutations. These mutations would often hit proto-oncogenes, yielding active oncogenes. The forced cell growth would collapse the time between mutations, speeding up the process that allowed a cell to acquire multiple mutated oncogenes.

So the theme first popularized by Ames required more subtlety than scientists initially appreciated. Mutagens were indeed carcinogenic, but other agents could be carcinogenic as well by favoring cell proliferation. Working hand in hand with mutagens, these growth-promoting agents, called "promoters," could hasten the process that leads to many kinds of human cancers.

....

CHAPTER 7

BRAKE LININGS: THE DISCOVERY OF TUMOR SUPPRESSOR GENES

The 1982 discovery of the *ras* point mutation held great appeal for molecular biologists, whose goal was invariably to reduce complex biological processes down to simple underlying mechanisms. They liked the notion that the development of cancer depended on nothing more than a single mutation in the genome of a normal cell. But within a year, with the discovery of collaborating oncogenes, the number of mutations that were apparently required to make a tumor crept up to two. Even this number held great appeal for them. Two mutant genes still represented a manageable level of complexity. Then even this number came under attack. By the mid-1980s, it became increasingly apparent that the number of mutations accumulated by most kinds of tumors during their development was far more than two. The evidence from epidemiology, remember, suggested that at least half a dozen steps were required to make a cancer; many scientists speculated that each of these steps involved the creation of a mutant gene in cells evolving toward a fully malignant state.

This realization triggered a search for the multiple mutant oncogenes that were predicted to be present in the genomes of human cancer cells. The researchers who began hunting for these genes came up with a surprise, a great disappointment. Large cohorts of mutant oncogenes cohabiting the genomes of the same human tumor cell simply could not be found. Some tumors were found to carry *ras* oncogenes, others *myc* or N-*myc* or *erb*B2. But very few carried even two oncogenes simultaneously. Something was very wrong. The notion that a cancer developed through the successive activation of a series of oncogenes had lost its link to reality.

There were two ways out of this quandary. Perhaps tumors did not really carry multiple mutated genes, contrary to substantial though indirect evidence. Alternatively, perhaps cancer cells really did carry half a dozen or more mutant genes, but most of these were unrelated to oncogenes. These other hypothetical genes might play equally important roles in forming human tumors. If this were really so, then the gene searchers had been looking under the wrong lamp post. Their narrow focus on oncogenes as the genes capable of causing cancer might have been misguided.

By the mid-1980s, mutant genes very different from oncogenes were finally found in human tumor DNA. The newcomers came to be called "tumor suppressor genes." Their discovery filled a major gap in the puzzle of how human tumors arise. This new class of genes was uncovered through experiments that were far removed from the work on viruses, gene cloning, and gene transfer—the kinds of experiments that had created the explosion of interest in oncogenes in the decade since 1975.

This other line of work made use of an odd experimental procedure termed "cell hybridization." Those who practiced this technique, most prominently Henry Harris at Oxford University, would take groups of cells growing on the bottom of a petri dish and force them to fuse together. These fusions—in effect, cell-to-cell matings—allowed Harris, and later others, to uncover fundamental truths about the behavior of genes operating inside cancer cells, including the discovery of these new cancer genes.

Long before these cell fusions were begun in the mid-1970s, geneticists had experimented with matings between whole organisms. As described earlier, the first systematic study of mating genetics was carried out in the 1860s by the Austrian monk Gregor Mendel, who hybridized different strains of pea plants. His work was forgotten for a generation, then rediscovered in 1900. It formed the foundation of modern genetics and led to the notion that biological information is transmitted in the discrete packets that came to be called genes.

The explosive growth of genetics in the twentieth century showed that all organisms, including even simple single-cell organisms such as bacteria and baker's yeast, rely on genes as the templates for making progeny. Also, virtually all organisms, ranging from bacteria to humans, were found to have evolved elaborate mating mechanisms. In every case, the same underlying motive was apparent: Mating enabled the exchange and mixing of genes between members of a species. Since all species comprised populations of genetically heterogeneous individuals, mating afforded the opportunity of testing novel combinations of genes. Novel gene combinations might yield offspring that were more fit than their parents. That increased fitness, in turn, powered the engine of evolution.

Controlled mating of genetically distinct individuals became a powerful tool for studying the behavior of genes—in particular, how the genes carried by one parent in a mating blended with those of its partner. While bacteria and yeast cells were found to mate with one another, cells prepared from mammalian tissues lacked that ability. The only matings naturally allowed between mammalian cells involved the fusion of sperm and egg. These facts prevented researchers from observing the outcome of mating dissimilar kinds of cells—bone cells from one person with bone cells from another, or bone cells with muscle cells of the same person.

Harris wanted to circumvent these limitations imposed by nature. So he began forcing animal cells to fuse with one another in the confines of a petri dish. These cell-to-cell fusions, though highly artificial, offered a way of mating cells of different origins with one another. The fusion technique used by Harris depended on the ability of certain virus particles to cause the outer membrane of one cell to fuse with that of another sitting nearby in the dish. The result was a single shared outer membrane which enclosed the nuclei of the two parent cells. Soon the nuclei would fuse, pooling their genes into a single common nucleus.

Under some conditions, these fusions could involve dozens of cells simultaneously, yielding enormous cells that were far too large and unwieldy to grow and divide. Far more interesting were the fusions involving only pairs of cells. Such two-cell hybrids could grow and divide, transmitting to offspring the pooled genes originating from the two parent cells.

These two-cell matings, like most marriages, were interesting only when the two partners were quite different from one another. As in much of genetics, the interest came in trying to predict the characteristics of offspring. Would the genes of one or the other be especially influential? Human genetics gave rise to similar issues: Will little Johnny have his father's eyes or his mother's? His red hair or her brown hair?

The often unpredictable outcomes reflect a struggle between the genes donated by the two parents. In the case of whole organisms, the offspring of matings, whether they be yeasts or humans, carry two copies of genes that template specific traits. These two gene copies may carry conflicting signals. Johnny may inherit one brown-eye gene and one blue-eye gene from his parents. The question is which gene copy will ultimately determine the appearance of his eyes.

The winners are usually termed "dominant" versions of genes, the losers "recessives." Dominant versions are often more potent in affecting cell metabolism. For example, a dominant version of an eye-color gene may specify the ability to manufacture an eye pigment, while the recessive version may simply lack the pigment-making ability.

With all this in mind, Harris began to fuse different combinations of human and rodent cells to see how their genes would blend with one another. The most provocative forced-cell marriages were between normal cells and cancer cells. He would grow mixtures of these cells together in a petri dish, fuse them pairwise, and then study how the normal-cancer hybrid cells behaved.

The outcome of these hybridizations seemed obvious. Cancer is a dominating force in the body, and tumor cells invariably grow more vigorously than their normal counterparts. Therefore, if a cancer cell was fused to a normal cell, the potent genes in the cancer cell would dominate over the weaker genes carried by its normal partner. The hybrid cell carrying both sets of genes should, by this logic, behave like the cancerous parent. Among other things, this hybrid cell should be able to seed tumors when injected into a mouse or rat.

But Harris found the exact opposite. His normal-cancer hybrids invariably lacked the ability to seed tumors. The prevailing preconception was wrong by exactly 180°. A normal cell's growth genes were dominant; cancer-causing genes were recessive.

There was only one logical way to explain Harris's bizarre result. Normal cells seemed to possess genes to program normal cell growth. The tumor cells, in contrast, must have discarded these genes during their progression to cancer and were therefore not influenced by these genes' growth-normalizing properties. Following the cell marriages that Harris arranged in the petri dish, these normalizing genes, donated by the normal partner cells, could re-impose their will on the cancer cells, forcing their growth back into line.

This thinking could be carried further. The genes present in the normal cells seemed to be slowing down growth. They acted, in effect, like brakes that allowed cells to counteract any tendencies to lurch forward into runaway growth. Cancer cells, having lost these genes, had lost their braking mechanism. Once the braking mechanism was reinstalled in the cancer cells by the cell hybridization trick, the cancer cells' forward momentum ground to a halt. Now these runaways had a means of controlling the uncontrollable—their drive to grow without limit.

None of this fit with the prevailing view that cancer genes acted dominantly. That view was conditioned by a decade's worth of research with oncogenes. When mutant activated oncogenes were introduced into cells carrying normal proto-oncogene versions, the oncogenes invariably dictated the outcome. They forced uncontrolled growth, overruling their normal counterparts. This implied that the proto-oncogenes, acting as recessive gene copies, functioned to promote normal, well-controlled cell proliferation; their mutant oncogenic versions were hyperactive, acted dominantly, and drove the continuous, unrelenting multiplication associated with cancer cells.

So Harris's growth-normalizing genes, which functioned so differently from proto-oncogenes and oncogenes, required a new designation. They came to be called tumor suppressor genes, a reflection of their behavior in the cell fusions. Hyperactive, dominant versions of proto-oncogenes and inactive, recessive versions of tumor suppressor genes both seemed to contribute to cancer.

It would take years before tumor suppressor genes were isolated by gene cloning, but the evidence pointing to their existence was undeniable. These tumor suppressors needed to be taken into account by all those trying to understand the genetic basis of cancer.

Now there were two groups of genetic actors on the cancer stage, each specifying a distinct part of the machinery that governed cell growth. The proto-oncogenes operated like the accelerator pedal in a car; mutant oncogenic versions of these genes seemed to result in pedals that were stuck to the floor. Conversely, the tumor suppressor genes worked like brakes. As normal cells developed into cancer cells, they might shed or inactivate these tumor suppressor genes, resulting in defective braking mechanisms. Runaway cell growth seemed to be explainable by either mechanism.

The existence of two diametrically opposite explanations of cancer formation demanded some resolution. Did some kinds of tumor cells rely on one mechanism to achieve malignant growth while others used the alternative mechanism? Or did both mechanisms operate together within cancer cells? Perhaps stuck accelerators and faulty brakes conspired to create cancerous growth.

The answers to these questions did not fall in place immediately. But the discovery of tumor suppressor genes did open the door to another aspect of human cancer—its heritability. Cancer often runs in families, and these genes provided an explanation for the origin of many kinds of familial cancer.

CANCER IN THE EYE

Harris's work indicated that the loss of a tumor suppressor gene played a critical role in launching some cancers. Once a cell rid itself of the inhibitory influence of such a gene, its growth program could lurch ahead. Without good brakes, there would be nothing to keep the car from racing out of control.

There are many ways a cell can deactivate or jettison a gene. Almost all involve mutation of the DNA sequences that constitute the gene. Often, long stretches of DNA bases are deleted from the midst of a gene. Occasionally, a whole region of a chromosome encompassing many genes may be thrown overboard.

But the most convenient and therefore most frequently used way for a cell to rid itself of a gene is more subtle. Most often, a gene will suffer only a single base change—a point mutation—in one of its sequences. Such a subtle change may have deadly consequences if it strikes a critical sequence in the gene. Point mutations may insert inappropriate punctuation marks into the middle of a gene; because these marks normally signal the end of a gene, they may terminate the read-out of the gene prematurely, causing truncation of the protein specified by this gene. Other times, the protein product of the gene may suffer some change in its string of amino acids that renders it nonfunctional. The result of all these mutations, large and small, will be the same: The cell will lose the services of the mutated gene.

In reality, losing the services of a tumor suppressor gene is more complicated than implied here. Almost all genes in our cells are present in two redundant copies, one deriving originally from our mother's genes, the other from our father. In the case of tumor suppressor genes, this two-copy system offers a measure of protection to the cell. If one copy of a suppressor gene is accidentally lost, the remaining gene copy serves as a perfectly adequate backup. Almost always, half a brake lining is as good as a whole one in slowing down cell growth.

This redundancy represents a general mechanism for preventing the onset of cancers in the body. If losing one suppressor gene copy is unlikely, losing both becomes extremely unlikely. In particular, the chance of losing any single gene through mutational inactivation is often in the range of one in a million per cell generation. The risk of losing both gene copies would therefore seem to be only about one in a million million per cell generation. However, the actual risk is higher—more like one in a billion—because of complex genetic mechanisms, some of which are discussed below. Still, the chance of a cell inadvertently losing both copies of these important growth-controlling genes is very low, creating an effective barrier to runaway cell growth.

The dynamics of the one-two punches that result in elimination of tumor suppressor genes are critical to the formation of many types of tumors. We first learned about these dynamics from study of a rare eye tumor, retinoblastoma. This tumor is seen in only one in twenty thousand children and occurs only up to the age of six or seven. In the United States, where more than half a million people die of cancer each year, the number of new retinoblastomas encountered annually is barely more than two hundred. These rare tumors seem to arise from cells in the embryonic retina that are usually destined to become photoreceptors—the rods and cones that sense light and respond by sending electrical signals through the optic nerve into the brain.

Retinoblastomas are classified in two categories. In sporadic forms of the disease, afflicted children have no close relatives who have previously contracted this cancer. In the familial form, this otherwise rare tumor can be documented to afflict more than one family member, often through several generations.

In 1971, Alfred Knudson, a pediatrician working in Texas, proposed a theory that united the two forms of retinoblastoma under a single genetic umbrella. He argued that two gene mutations were required to strike a retinal cell before it spawned a retinoblastoma. In the sporadic form of the disease, these two mutations would occur one after another, either during embryonic development or shortly after birth, in one of the cells of the retina. The doubly mutated retinal cell would then begin to grow uncontrollably.

In the familial form of the disease, Knudson argued, these critical mutations arose quite differently. One of the two mutations would already be present in the fertilized egg from which a child developed. Such a mutation might be inherited from a similarly afflicted parent, or might even arise during the formation of sperm or egg. This mutation would then be transmitted to all of the cells of the developing embryo. Accordingly, a mutant copy of this gene would be present in all the cells of a newborn, including, most importantly, the cells of its retina. Thereafter, any one of these retinal cells would need only a single additional mutation to arrive at the doubly mutated state necessary to trigger eye cancer.

Recall that somatic mutations strike at the genomes of all cells outside of the gonads. Because mutations are rare events, the probability of two somatic mutations converging on a single retinal cell is extremely rare. Indeed, sporadic retinoblastoma affects only one child in forty thousand; children with this form of disease invariably carry only a single retinal tumor.

In the familial form of retinoblastoma, in contrast, only a single rare somatic mutation is required to trigger the explosive outgrowth of a tumor. Since the number of target cells in the retina seems to be large (more than ten million) and the chance of a single mutation is one in a million per cell, children inheriting a mutant gene and associated predisposition to retinoblastoma often have multiple tumors in both eyes. In effect, each of their retinal cells is poised on the edge of cliff, requiring only a single somatic mutation to push it over.

By the mid-1980s, the nature of these mutations and the genes they affected had become clear. The two target genes were the paired copies of a gene that sits on the thirteenth human chromosome and which, because of its disease association, came to be called Rb. Each of the mutations predicted by Knudson served to knock out one of the Rb gene copies. When only a single gene copy was deactivated, a retinal cell could still rely on the surviving partner gene and would continue to grow in a fully normal fashion. But when both Rb gene copies were lost, control of proliferation would be totally disrupted—the cell would have lost its brakes.

The Rb gene had all of the properties of the tumor suppressor genes predicted by Harris's cell fusion experiments. It was present in the genomes of normal cells and absent or functionally inactive in the genomes of tumor cells. But now there were additional insights that could be layered onto those coming from Harris's earlier work. First, the loss of tumor suppressor gene function was a two-step process, involving the successive elimination of two gene copies. Second, defective versions of a tumor suppressor gene could be passed from parent to offspring via sperm or egg to create an inborn susceptibility to cancer.

The DNA sequences composing the Rb gene were isolated by gene cloning in 1986 in a collaboration between my laboratory and that of Thaddeus Dryja. This cloning made it possible to gauge the full role that the Rb gene plays in the genesis of human cancers. At first, it seemed that this role was limited to causing the very rare eye tumors in children. Indeed, it seemed to be mutated in all these tumors. In addition, children who survived childhood familial retinoblastoma were known to be at increased risk for bone cancers (osteosarcomas) as adolescents; loss of Rb gene function was demonstrated in these tumors as well.

By the late 1980s, use of the recently cloned Rb gene revealed that more than a third of bladder cancers and a smaller proportion (about 10 percent) of breast cancers also lost Rb gene copies, in both cases through somatic mutations that occurred in these two target organs. Most surprising was the genetic analysis of small-cell lung carcinomas, a tumor that represents one of cigarette smokers' most common routes to the grave. Almost all of these tumors were found to have jettisoned both Rb gene copies, one after the other, during tumor formation.

We came to realize that the Rb gene plays a far broader role in cancer causation than first imagined from its initial association with a rare childhood tumor. This catalog of Rb-associated cancers creates a major puzzle: What common trait unites the cells in the various affected organs throughout the body? The Rb gene functions in all cells of the body to brake their growth. Why is this peculiar constellation of tissues particularly susceptible to becoming cancerous following loss of the Rb gene? The answers may not be known for many years.

LOSS OF DIVERSITY
....

We now know of more than a dozen tumor suppressor genes, the Rb gene being only the first on the list. Finding them has not been easy. The existence of these genes becomes apparent only when they are absent. How can one discover genes that behave like ghosts, influencing cells from behind some dark curtain?

Some of these genes are associated with familial malignancies like retinoblastoma; like the Rb gene, they can be passed in mutant defective versions through the germ line. Other tumor suppressor genes are not associated with inherited susceptibility to cancer. These other suppressor genes are lost through somatic mutations that strike locally in one or another target organ and proceed to eliminate, one after the other, the two copies of these genes.

A clever genetic trick has made it possible to track down many of these genes. The trick depends on the details of the genetic mechanism by which the two copies of these tumor suppressor genes are lost as tumors develop. The most straightforward process is a scheme in which one gene copy is lost at a frequency of one in a million per cell generation. Later on, another rare one-in-a-million event strikes again in the same cell or one of its direct descendants, knocking out the surviving gene copy. After both gene copies are lost, run-away cell growth follows. As mentioned earlier, the probability of two such events striking the same cell (or small cohort of cells) is the product of the probability of each occurring separately, about one in a million million per cell generation. This event is so improbable that it occurs only rarely in the body during a normal life span.

Tumor cells usually resort to a more expedient way of eliminating the second copy of a tumor suppressor gene. Their strategy depends on the fact that the two partners in a human chromosome pair (such as the two thirteenth chromosomes, each of which carries an Rb gene copy) often line up next to one another in parallel array, look each other over, compare their respective DNA sequences, and then swap genetic information. One frequent result is that a gene sequence present on one chromosome will now replace the corresponding sequence carried by its partner. Before this information transfer, two distinct versions of a gene may have resided on the two paired chromosomes; afterward, one of these versions is lost, being replaced by a duplicated version of the gene originally present on the other chromosome. The result is two identical copies of a gene in a cell that previously carried two dissimilar versions.

This loss of genetic diversity in the cell is often called "loss of heterozygosity." The two copies of the gene are now rendered equal—they become homogenized. This homogenization of one or another gene occurs rather frequently, as often as once in every thousand cell divisions. For this reason, it represents a facile means by which the still-intact copy of a tumor suppressor gene may be lost. In short, the intact gene copy is discarded and replaced by a duplicated copy of the already mutated, defective version of the gene. The overall probability of this occurring is one in a million (inactivation of the first gene copy) multiplied by one in a thousand (duplication of the inactivated gene copy and loss of the active one), which yields altogether a frequency of one in a billion per cell generation.

Precancerous tumor cells on their way to becoming actively malignant will often use this trick to eliminate both copies of a tumor suppressor gene that has been holding back their growth. They will first mutate to inactivity one copy of the gene and then eliminate its partner through this loss-of-heterozygosity homogenization. Importantly, the chromosomal swaps that generate this homogenization often involve large regions of the chromosome surrounding the tumor suppressor gene, not just the gene itself. Hundreds of genes residing to the left and the right of the tumor suppressor gene on a chromosome will also become homogenized.

Of course, the homogenization of neighboring gene copies is irrelevant for the growth of the developing tumor cells. These neighboring genes are nothing more than innocent bystanders. It is really only the tumor suppressor gene that the tumor cell is intent on eliminating through use of the homogenization trick.

The fate of these neighboring genes offers an entree for the geneticist intent on locating and isolating novel tumor suppressor genes. A large collection of randomly chosen genes scattered throughout the chromosomes of a tumor cell can be analyzed for loss of heterozygosity. The geneticist looks for genes that are present in two discordant versions in the DNA of normal cells but are represented in two identical versions in the same person's cancer cells. Such loss of diversity suggests that this gene, whatever its identity, lies near on the chromosome to a tumor suppressor gene that was the actual target of homogenization during tumor cell development.

With this logic in mind, geneticists have launched hundreds of searches in tumor cell genomes for chromosomal regions that are repeatedly homogenized during tumor progression. These regions are under suspicion of harboring tumor suppressor genes. Once such regions are located, gene cloning techniques can be brought to bear to find and then isolate the suspected culprit genes.

To date these searches have netted more than a dozen genes for the gene cloners. The chromosomal region around the *Apc* gene is homogenized during the evolution of almost all colon carcinomas. The region around *NF*-1 loses diversity during the genesis of neurofibromas. The *WT*-1 chromosomal region suffers the same fate in some childhood kidney cancers while that carrying *VHL* is lost in many adult cases of this disease. The *p16^{INK4A}* gene loses heterozygosity during the development of a variety of tumors.

This roster leaves the impression that the human genome carries many tumor suppressor genes. Estimates range as high as three or four dozen, but these numbers are very imprecise. These gene discoveries, like the earlier one that led to the cloning of the Rb gene, create an intriguing puzzle, still unsolved: While most of these genes operate in many cell types throughout the body, loss of most of these usually has a strong impact on the growth regulation of certain tissue types but not others.

But certain genes stand as strong exceptions to this pattern of tissue-specific targeting. The *p53* tumor suppressor gene plays a prominent role in a wide diversity of cancers, being present in mutant form in as many as 60 percent of all human cancers. Mutant versions of the *p53* gene can also be transmitted from parent to offspring, who will then be susceptible to a wide range of carcinomas and sarcomas throughout life.

The searches for new tumor suppressor genes remain laborious. Each gene requires many man-years of effort. After all, the discovery of loss of heterozygosity in the chromosomes of a certain tumor cell type only suggests a starting point for a molecular hunting party that must still sweep through several million base pairs of DNA in order to find a culprit gene.

As the Human Genome Project progresses in its attempts to catalog and map all human genes, the discovery of new tumor suppressor genes will be enormously simplified. Soon the years of effort needed to find even one of these genes will be compressed into several months, and many missing pieces will be filled into the genetic puzzle of cancer. With these genes in hand, we will be able to write definitive life histories of many tumors in terms of the mutant oncogenes and tumor suppressor genes that these tumors have accumulated on their road to malignancy.

....

CHAPTER 8

GUTS:
A CASE STUDY IN
CANCER DEVELOPMENT

The human bowel provides a particularly fertile ground for cancer. It was not always so in our history, or at least colon cancer was not a common cause of death until recently. Two things changed in modern times. We now live much longer than we used to. By the middle of the twentieth century, many of us began to live for seventy or eighty years, an age when this disease strikes frequently. A hundred years earlier, relatively few survived long enough to confront it. Our diet also changed from one heavy in grains and vegetables to fare that increasingly emphasized meat and large amounts of fat. The effects of diet are apparent from epidemiology: There are regions of Africa whose inhabitants follow a diet based almost exclusively on vegetables and grains. These people run a risk of colon cancer that is less than one-tenth of that seen in the West.

By mid-century, increased longevity and changing diet had conspired to create enormous numbers of colon cancers in the American population. For those interested in understanding the development of a specific type of human cancer, the colon became an extremely attractive site to study. In many other organ sites, tumors appeared by the hundreds or several thousands each year in the United States. The colon offered an embarrassment of riches—more than a hundred thousand new cases diagnosed annually.

The gut offered another advantage as well. Unlike most other internal organs that are frequently struck by cancer, the colon is relatively accessible. A colonoscope—a flexible optic tube inserted through the rectum—affords a direct view of the cells lining the bowel cavity. By the late 1980s, millions of surveys of human colons, normal and malignant, had generated a wealth of data on how things can go wrong in this complex tissue.

Cells that form the normal colonic epithelium—the layer of cells lining the walls of the large intestine—are normally turned over rapidly. Typically, these epithelial cells are formed, mature, and then are sloughed off into the cavity of the bowel within a two- or three-day period. This rapid turnover suggests that these cells have only a short useful lifetime, presumably because they are vulnerable to attack by the contents of the intestine—the products of digestion and the large populations of bacteria that inhabit our gut. In effect, the wall of the intestine continually retires cells from the line of fire after a short period of service and replaces them with fresh recruits. This prevents the accumulation of damaged, defective cells, including those that may have sustained mutations in their growth-controlling genes.

Despite this unceasing turnover, the lining as a whole usually remains remarkably constant and well organized. The view through the colonoscope shows that its overall architecture is usually maintained faithfully throughout life. But in some individuals, maintenance breaks down and abnormal tissue structures appear. These aberrant architectures range from an excess of normal-looking cells (hyperplasia), to clumps of cells that have taken on some but not all of the attributes of cancer cells (dysplasia), to large masses of dysplastic cells, variously termed adenomas or polyps, which often protrude into the bowel cavity.

The most extreme changes are associated with obviously malignant growths (neoplasia). Like all malignancies of epithelial tissues, these are considered carcinomas, and they take on various guises. Some are relatively localized, while others invade the muscle wall underlying the intestinal lining and, in more advanced stages, send out offshoots that colonize neighboring organs, including most notably the liver.

This sequence of increasing disorder represents more than a convenient way of arranging complex descriptive information. It hints at an important biological reality: The development of a colon carcinoma may encompass a series of steps in which normal cells and tissues pass through stages of increasing abnormality, beginning as fully normal and ending as highly malignant.

Such a sequence of incremental changes seems to echo a theme we encountered earlier that portrayed the development of a tumor as the end product of a long, multistep sequence of genetic events. Perhaps the various premalignant growths attached to the bowel wall represent intermediate stopping points on the road from the fully normal to the fully malignant. If true, this picture implies that fully malignant tumors can arise only from already abnormal, premalignant growths, never directly from normal tissues.

Although this scheme is attractive, like many proposed explanations of cancer's origins, it may be little more than simplistic thinking driven by a desire to explain a complex phenomenon by invoking a simple underlying mechanism. In fact, the catalog of different growths in the colon is also compatible with another scheme, in which the normal colonic lining can convert itself in single large steps into a whole variety of growths, some moderately abnormal, others frankly malignant. Maybe normal cells occasionally leapfrog over intermediate states, becoming cancerous in one giant step. The pictures taken through the colonoscope fail to tell us how these normal and abnormal tissues are related to one another.

But analysis of the mutant genes carried by these growths has shed much light on these relationships. Bert Vogelstein of Johns Hopkins Medical School in Baltimore began such gene work in the late 1980s. He analyzed biopsy samples from several types of intestinal growths, looking for obvious genetic abnormalities. The information he assembled provided strong support for the idea that tumors progress from normality to malignancy in small steps; as they advance through this progression, colonic cells accumulate increasing numbers of mutant genes in their genomes.

Vogelstein discovered that the fifth, seventeenth, and eighteenth chromosomes frequently suffer loss of diversity (heterozygosity) as growths progress toward becoming malignant tumors. This observation pointed to the existence of tumor suppressor genes on each of these chromosomes, elimination of which seemed to be vital to the progression of carcinogenesis in the colon.

Both copies of the *Apc* tumor suppressor gene on chromosome 5 were already mutant in cells of the very early, slightly abnormally growing polyps. As the polyps became more advanced, an additional mutant gene was detectable in their DNA—the *ras* oncogene. Still more advanced polyps poised on the brink of becoming malignant seemed to have lost a suppressor gene that Vogelstein termed *DCC*. Finally, colon carcinoma cells may carry mutant versions of these three genes plus altered versions of the *p53* tumor suppressor gene.

Observations such as these prove that cancer development is a complex process of multiple steps. This strengthens the notion that tumors develop through a Darwinian process involving repeated rounds of mutation followed by selection. Also, the notion that a full-fledged cancer can develop in a single step from normal tissue is rendered highly unlikely.

Not all tumors followed exactly this sequence of mutations. Other still-to-be-discovered genes and gene mutations can participate as stand-ins for the genes cited above. But this caveat does not weaken the major lesson. Tumor formation really does depend on a succession of mutations that collaborate to create the aggressive growth associated with late-stage cancers.

Significantly, the sequence of mutations involves both tumor suppressor genes and at least one oncogene. The suppressor genes become deactivated while the oncogene undergoes hyperactivation. The automobile metaphor comes to mind once again: The growth of cancer cells benefits from the simultaneous flooring of the accelerator and loss of brakes.

The model of oncogene collaboration described earlier now requires revision. Recall that single oncogenes acting alone cannot transform fully normal cells into tumor cells, while various pairs of oncogenes (such as *ras* and *myc*) can indeed cooperate with one another to induce cell transformation. This suggested that cells accumulate several mutant oncogenes on their way to becoming cancerous. In reality, few human tumors carry multiple mutated oncogenes; the colon cancer example—activation of an oncogene (such as *ras*) together with the inactivation of tumor suppressor genes (*Apc, DCC,* and *p53*)—is much more typical. Hence, the activation of oncogenes cooperates with the inactivation of tumor suppressor genes to create cancer.

CARPETING
....

Like retinoblastoma, colon cancer runs in families. About 1 percent of colon cancers in the American population are due to an inherited condition known as familial polyposis. In affected families, a mutant gene passed from generation to generation causes members to develop multiple polyps in the colon. The polyps may number in the thousands and appear to carpet the walls of the large intestine.

These polyps are benign, localized growths. Still, each one has a small but significant probability of developing into a malignant tumor. In an individual with familial polyposis, the number of polyps is so large that it is almost inevitable that one of them, sooner or later, will undergo this conversion and erupt into a life-threatening colon carcinoma.

The genetics of transmitting susceptibility to polyposis from parent to offspring closely follows the pattern of retinoblastoma. Once again, a mutant, defective version of a tumor suppressor gene is passed through sperm or egg. The child inheriting this gene is destined to sustain a tumor in a specific target organ, in this instance the colon. And as in retinoblastoma, at some point a cell in the target organ will lose its surviving intact copy of the tumor suppressor gene and launch forth into unconstrained proliferation.

The inherited mutant gene here is one that we have already encountered—*Apc.* In sporadic carcinomas, described in the last section, the deactivation of one copy of this gene is the first step in the multistep process of colon tumor progression. Subsequently, the other *Apc* gene copy is discarded by cells on their way to becoming cancerous.

Individuals inheriting a defective copy of *Apc* have already taken the initial step in this multistep process. All of their colonic cells already carry a mutant gene copy and can, as a consequence, proceed directly to the next step—elimination of the remaining intact *Apc* gene. In these people, the process of polyp formation and ultimately carcinoma formation is greatly accelerated.

So familial polyposis and familial retinoblastoma provide a great simplification in our thinking about sporadic and familial cancers. We can now unify familial malignancies, which are preordained from conception, with the more common sporadic malignancies, which arise because of random genetic accidents occurring throughout life. A common set of genes is involved. Their presence in mutant form in the germ line (sperm or egg) generates a congenital predisposition to cancer; the same genes, when mutated by random processes in the cells of a target organ, yield the unpredictable tumors that represent more than 90 percent of the cancer burden in most human populations.

··

CHAPTER 9

DEFENDERS OF THE TEXT:
DNA REPAIR AND ITS BREAKDOWN

As we saw in the last chapter, defects in two classes of growth-controlling genes—oncogenes and tumor suppressor genes—contribute to the inception and subsequent development of tumors in the colon. The joint involvement of these two classes of genes is essential for the generation of a variety of tumors; to date it has been demonstrated in bladder, lung, brain, and breast tumors. Over the next decade, this principle will likely be extended to tumors arising in virtually all tissues in the body. Different types of tumor cells will undoubtedly contain mutations in distinct sets of genes. We already know that the tumor suppressor genes and oncogenes participating in cancer formation in the breast are in large part different from those encountered in colon cancer. But in all cases, the general theme survives: The malignant growth of the human cancer cell derives from the activation of oncogenes and the deactivation of tumor suppressor genes.

Recently, it has become clear that yet other kinds of genes can also play major parts in the creation of cancer. In their normal incarnations, these other genes are not responsible for governing cell proliferation. Their job inside the cell is quite different: They work directly or indirectly to maintain the integrity of the cell's DNA. Their failure to do so results in large numbers of mutations accumulating in genes throughout the cell genome, including notably the growth-controlling genes that we encountered earlier. Because the rate of mutation of these growth-controlling genes is accelerated, the overall process of cancer formation is quickened, yielding large increases in the numbers of tumors arising in a lifetime.

The genetic text stored in the DNA sequences of the human cell is always susceptible to corruption. Many chemical carcinogens introduced into the body through food or tobacco consumption eventually enter our cells. Once inside, many of these chemicals attack the cells' DNA molecules. The great majority of dietary mutagens are likely to be natural components of our food rather than man-made contaminants. Ames has documented dozens of natural foodstuffs, from brewed coffee to celery stalks to bean sprouts, that contain high concentrations of natural, highly potent mutagens.

Moreover, as Ames and others have described, the cell's normal energy metabolism releases millions of highly reactive molecular by-products every day. Many of these are oxidants and "free radicals," the latter bearing unpaired, highly reactive electrons. Like environmental mutagens, these indigenous molecules can chemically alter a variety of molecules in the cell including the cell's DNA. Once again, the information content of the DNA may be affected.

Most of these reactive molecules are intercepted and neutralized by an elaborate phalanx of protective molecules that the cell maintains to ward off their attacks. Among these molecules are natural anti-oxidants such as vitamin C. The cell also manufactures a variety of enzymes to neutralize and detoxify harmful molecules before they have a chance to wreak genetic havoc.

Some humans make these detoxifying enzymes at high levels, while others produce them at much lower levels. These differences, which are heritable, offer us the opportunity to understand the roles of these enzymes in protecting cells against attack by various carcinogens. For example, do individuals with low levels of these protective enzymes contract cancer more often than those having high levels?

In fact, some striking differences have been uncovered. Smokers who make low levels of the NAT enzyme (N-acetyl transferase) run as much as two and a half times the risk of bladder cancer compared to smokers having high levels of the NAT enzyme. Low levels of a second detoxifying enzyme, GSTM1 (glutathione-S-transferase M1), result in a threefold increase in the risk of lung cancer. These findings suggest that we may one day be able to calculate smokers' risk of cancer based on their lifelong cigarette consumption and their levels of detoxifying enzymes.

Some mutagens succeed in penetrating this complex array of protective devices. Having escaped inactivation, these mutagens may proceed to react with and thereby damage the DNA molecules carried in the cell's chromosomes. Every human cell sustains thousands of such mutagenic attacks every day. But in spite of this barrage, the cell's DNA emerges at the end of the day relatively unscathed. This discrepancy demands explanation.

Close scrutiny of the machinery used by the cell to copy its DNA molecules reveals a similar discrepancy. The process by which cells replicate DNA in preparation for cell division is prone to error. Immediately after DNA polymerase—the enzyme responsible for DNA replication—has copied a stretch of DNA, as many as one in every thousand bases of the newly made DNA strand may be incorrect, having been mistakenly inserted by the polymerase. But as before, the actual rate at which mutations accumulate in the DNA is much lower. Somehow, the vast majority of these initial copying mistakes are not perpetuated in the DNA.

The actual numbers are remarkably low: By the time a cell has completed the entire process of DNA replication, fewer than one DNA base in a million appears to have been miscopied. This low mutation rate provides a striking testimonial to the efficiency of a cellular copyediting apparatus designed to detect miscopied bases in the DNA and remove them from the double helix. The removed bases are then replaced with new ones that regenerate the proper sequence in the DNA, erasing all traces of the previous miscopying. A similar apparatus operates to detect and cut out DNA bases that have been attacked and altered by chemical mutagens. The processes that restore the genetic text are termed "DNA repair."

So the rock-solid stability of the cell's genetic data base is a mirage. The constancy of our genome is the result of a high-wire balancing act, a permanent struggle in which an ever-vigilant repair apparatus continuously fights off genetic chaos.

This dynamic has direct consequences for tumor formation: If the DNA repair process fails, large numbers of altered bases will accumulate in a cell's DNA. This means that the rate at which mutations accumulate is influenced by at least three distinct processes: damage inflicted on the DNA by mutagens of foreign or internal origin, mistakes made during DNA copying, and defects in the DNA repair machinery responsible for erasing the damage created by mutagens or miscopying. Since mutation is the engine that drives tumor progression, all three of these processes are likely to be involved in one way or another in causing cancer.

We now know that several kinds of familial cancers are caused by inherited defects in DNA repair. The cellular apparatus responsible for repairing DNA consists of a large cohort of proteins. Some of these are specialized to recognize damaged DNA segments, others to cut them out, yet others to replace the excised segments with new ones that restore the proper base sequence. A defect in one of the genes that specify the structures of these proteins can propel the process of cancer formation at very high rates.

A dramatic case in point comes from a form of inherited colon cancer that is four or five times more common than familial polyposis. An individual suffering from hereditary non-polyposis colon cancer (HNPCC) inherits a defective version of one of four distinct genes specifying critical DNA repair proteins. These four proteins are all essential components of the machinery that is specialized to repair mistakes occurring during the process of DNA replication. As mentioned above, each time a cell replicates its DNA, a number of copying errors will rapidly be erased by replacing miscopied bases with the appropriate ones. In the cells of a person suffering from HNPCC, however, many of these copying mistakes will remain uncorrected, and then will be passed on as mutations to the daughter cells following cell division. As a consequence, over many cycles of growth and division, the cells of HNPCC patients accumulate mutations at an alarming rate.

In an HNPCC patient, all cells throughout the body appear to suffer from defective DNA repair. Yet despite this global defect, the major increased cancer risk is manifested largely in the gut and the endometrial lining of the uterus; other sites, affected to lesser extents, include the ovary and the bladder. The reasons for this peculiar constellation of cancers are still obscure.

The colon tumors that arise in HNPCC patients carry mutant oncogenes and tumor suppressor genes much like those seen in sporadic, nonfamilial cancers. The essential difference is the rate with which these genes undergo mutation. In the colonic cells of an HNPCC patient, the absence of a competent DNA repair apparatus ensures that the rate of mutation of these genes, and thus the overall rate of tumor development, is greatly accelerated.

Among the diverse DNA repair enzymes are those specialized to recognize the damage created by ultraviolet (UV) rays. This short-wave radiation from the sun or tanning lamps can create substantial damage in skin cells by striking DNA molecules and causing neighboring bases in DNA strands to fuse into bizarre double-base complexes. These base fusions later cause miscopying and thus the accumulation of mutations. The outcome may be basal cell or squamous cell skin cancer, both of which are highly curable, or melanoma, which often is not.

Skin cancer has been increasing in frequency in recent years. Rates of melanoma have been increasing about 4 percent each year for the past two decades. The cause of this increase is almost certainly the great increase in sunbathing over the past three or four decades. Use of tanning parlors will almost certainly accelerate this trend. In spite of the best efforts of the DNA repair apparatus, many mutant skin cells accumulate in people who, through design or accident, have suffered heavy and repeated doses of UV light.

The rare disease xeroderma pigmentosum arises from an inherited defect in one or another of the ten or so genes responsible for the repair of UV-induced DNA lesions. The skin of these individuals is extremely sensitive to sunlight and will frequently sprout skin cancers. The survival of xeroderma pigmentosum patients depends on total avoidance of direct sunlight and frequent screening of their skin for potentially malignant growths.

Individuals who have inherited defective versions of another type of DNA repair gene, termed *ATM*, are extraordinarily vulnerable to ionizing radiation, or X-rays. This sensitivity is only one indication of a widespread defect in DNA repair, the tip of an iceberg; all their lives, these people accumulate mutations throughout their bodies at an accelerated rate.

This *ATM* gene defect may rear its head in several ways. A person inheriting two defective copies of this gene will be afflicted with the syndrome known as ataxia-telangiectasia. These people—numbering about one in fifty thousand individuals—pay a high price for their defective DNA repair. They often have unsteady posture, dilated blood vessels, immune deficiencies, premature aging, and a hundredfold increase in the risk of cancer.

Recent evidence also suggests that two genes implicated in familial breast and ovarian cancers—*BRCA*1 and *BRCA*2—specify proteins that are also involved in maintaining the integrity of the cell's DNA. As many as 10 percent of breast cancers in the American population may be due to inheritance of defective versions of one or the other of these genes. As with other inherited DNA repair defects, it is unclear why these two mutant genes preferentially target certain organs, in this case the breast and the ovary.

The full complexity of the DNA repair apparatus is still not understood. Equally unknown are the spectrum and frequency of defective repair genes carried in the human population. Only when these are thoroughly cataloged will we begin to appreciate the role of defective DNA repair in many types of human cancers.

Even more complex are the enzymes involved in neutralizing mutagens of foreign origin, such as those introduced into the body by cigarette smoking. Another decade may pass before we understand the contributions made by these various enzymes to the defense of the genome against chemical attack, and the price we pay when adequate levels of these enzymes are lacking in our cells.

CHAPTER 10

GUIDE PROTEINS OF THE CELL: THE MACHINERY THAT CONTROLS GROWTH

Knowing about mutant genes allows us to trace the roots of cancer back to discrete, identifiable changes in the central controlling molecule of the cell, its DNA. But in one sense, these genetic discoveries are sterile and uninformative. Genes are pure information, nothing more than mathematical abstractions. Studied in isolation, they tell us little about the real life of the cell. Moreover, the sequence of DNA bases that constitutes a gene usually reveals little about how this gene operates. So, even after we know that one or another gene is mutated during the development of a cancer, we still understand next to nothing about the mechanisms by which this mutant gene causes abnormal cell growth. Fortunately, molecular biology provides us with a useful train of logic that leads us toward an understanding of gene function. Genes instruct the cells around them to make specific proteins. It is the proteins that do the work of the genes. Proteins catalyze biochemical reactions or create elaborate physical structures. To understand how a gene works, we must know intimately how its protein functions.

This logic dictates that each of the oncogenes described earlier encodes the structure of a specific protein. Once synthesized under the close supervision of its controlling gene, the oncogene protein sallies forth and effects changes in the cell. The *src* gene makes a protein termed $pp60^{src}$, while *ras* makes as its product $p21^{ras}$. The long list of oncogenes is paralleled by a corresponding catalog of oncogene proteins, sometimes termed oncoproteins. Of course, tumor suppressor genes also regulate cell proliferation through their own cohort of encoded proteins. In the end, a deep understanding of the disease of cancer can only come from detailed insight into how these various proteins operate.

Before we confront oncoproteins directly, we must place them in some biological context. In particular, we must learn how their normal counterparts contribute to the life of a normal, healthy cell. Normal cell function provides the baseline from which to study the molecular aberrations of cancer.

In one sense, the roles played by the normal versions of oncoproteins are obvious: They help the normal cell regulate its growth. Unfortunately, this statement does not take us very far; it only restates the problem, and in a way that is not terribly useful. A more productive tack is suggested by the following question: Precisely how do normal cells know when to grow and when to hold back from growth?

At any moment, the great majority of cells in our body are in a quiescent state. Only in tissues that renew themselves constantly, such as the colonic epithelium, the bone marrow (which generates new blood cells), and the skin, does one find large numbers of cells actively growing and dividing.

These dramatic differences in the proliferation rates of tissues bring us back to our question: Precisely how do any of these cells know when they should grow? The issue becomes even more complicated in the case of embryonic development, where cell proliferation results in the creation of new, complex tissues rather than the maintenance of an existing tissue architecture.

Although each cell carries an extraordinarily elaborate data bank in its genes, these genes cannot provide the cell with some very critical pieces of information. Genes cannot tell a cell where it is in the body, how it arrived there, or whether the body requires it to grow. Genes can only tell the cell how it should respond to external signals, which must come from elsewhere—from other cells, nearby and distant in the body. Each cell in the body relies on a host of other cells to tell it where it is, how it got there, and what it should be doing. Among other information provided by its neighbors (nearby and distant) are the instructions that tell a cell when it should grow.

Complex organisms could not be organized otherwise. Cells exist in condominiums with other cells, forming tissues, organs, and ultimately whole organisms. The behavior of an individual cell in these communities must be dictated by the needs of the organism around it. Hence, each cell must be in close and constant contact with many other cells in the organism; these contacts form the network that binds this community together. While cells within a tissue are physically tethered to one another, they are tied together even more importantly by incessant chatter.

A normal tissue is thus a network of millions of cells in constant communication, passing information to one another about their respective needs. How does a malignant tissue fit into this pattern? What characterizes the behavior of a cancer cell that arises in the midst of a crowd of normal neighbors?

The cancer cell is a renegade. Unlike their normal counterparts, cancer cells disregard the needs of the community of cells around them. Cancer cells are only interested in their own proliferative advantage. They are selfish and very unsociable. Most important, unlike normal cells, they have learned to grow without any prompting from the community of cells around them.

Now the question of how normal cells control their proliferation can be restated in more focused terms. A normal cell has an absolute requirement for external prompting before it will undertake to grow and divide. In contrast, a cancer cell seems able to stimulate its own growth, rendering it independent of prompting from other cells.

How then do cells stimulate each other to grow? Once we know this, we can begin to understand how oncoproteins usurp this normal cell-to-cell signaling and make it unnecessary. What tells a normal cell that some of its neighbors have died and that, as a consequence, it should grow and divide to fill in the gaps in the ranks?

MESSENGERS OF GROWTH
. . . .

In principle, information regulating growth might be passed from one cell to another by electrical signals or small organic molecules. For various reasons, however, evolution has solved this problem in another way. In all complex, multicellular organisms, this information is conveyed by small, soluble protein molecules termed growth factors. A growth factor protein is released by one cell, moves through intercellular space, and ultimately impinges on its target—another cell. The target cell then responds by initiating a program of growth and division.

Some growth factors are released at one site in the body and travel great distances through the blood before they home in on appropriate target cells. But more often than not, growth factor molecules act over very short distances, being released by one cell and then impinging on a close neighbor. It is largely this short-range signaling that knits together the community of cells within a tissue.

The synthesis and release of growth factors is under extremely tight control. Inappropriate release might stimulate cells to begin proliferation at a time and place where growth would lead to disastrous distortion of normal tissue architecture. We know rather little about the mechanisms that determine how and when cells release growth factors. But there are vivid examples that provide some insight into this control.

When a wound occurs in a tissue, blood clots assemble to stanch the bleeding. Integral to clot formation are the blood platelets, which gather at the site of bleeding and aggregate to form a physical barrier that prevents further loss of blood. At the same time, these platelets release several growth factors—notably a factor called platelet-derived growth factor (PDGF)—that stimulate nearby connective tissue cells to grow. These connective tissue cells are the pioneers that begin reconstruction of the damaged tissue, thereby healing the wound.

Controlled growth factor release also occurs when tissues are deprived of adequate oxygen. Cells in such a tissue will respond by releasing vascular endothelial growth factor (VEGF). This factor stimulates other nearby cells that specialize in constructing blood vessels. As a consequence, small capillaries near the site of VEGF release begin to extend into the oxygen-starved tissue. Soon, this tissue is penetrated by an extensive network of capillaries that supplies it with much-needed oxygen.

The importance of growth factor stimulation to the lives of cells becomes even more striking when cells are removed from living tissues and cultured in petri dishes. The culture fluids in petri dishes contain nutrients—sugars, amino acids, and vitamins—that every cell requires for its normal metabolism. But this nutrition only suffices to keep cells alive. In the absence of any explicit signal to grow, normal cells will sit on the bottom of a petri dish indefinitely without undertaking growth and division.

Only when serum is added to the nutrient culture medium will normal cells begin to proliferate. The added serum contains growth factors, most prominently PDGF. Other serum factors, such as epidermal growth factor (EGF) and insulin-like growth factor (IGF), may collaborate with PDGF to induce cells in a dish to begin growing.

It is clear that the proliferation of normal cells is absolutely dependent on signals from external sources. These cells will never proliferate on the basis of their own internally generated decisions. In the jargon of the sociologists, normal cells are totally "other-directed." Their behavior is dictated entirely by the world around them.

Cancer cells seem to violate this rule. In a petri dish, many kinds of cancer cells can grow with little or no serum added to their culture medium. While normal cells will only grow well if 5–10 percent of their medium is derived from added serum, cancer cells may grow in 1 percent serum or even less. This shows that cancer cells are much less dependent on external signals for their growth. It seems as if cancer cells respond to their own internal growth-stimulating signals. The answers to understanding cancerous growth have come from following up this speculation.

CELLULAR ANTENNAE

A specific set of molecules enables each cell to sense growth factors in the space around it. The surfaces of all cells are studded with "receptors" that serve as their antennae. These receptors allow cells to detect growth factors swimming through the medium that surrounds them. After a receptor senses a growth factor, it will transmit information about this encounter through the outer membrane of the cell into its interior. It is this signal transmission across the membrane that informs the cell that an encounter has taken place.

Receptor molecules have a remarkable structure. They are long protein chains. One end protrudes into the extracellular space, the middle threads its way through the membrane of the cell, and the other end sticks into the cell interior. The extracellular portion is responsible for sensing the presence of growth factors; the intracellular portion is assigned the role of releasing biochemical signals into the cell following these encounters.

Each growth factor has its own private receptor. The EGF receptor is specialized in sensing EGF in the extracellular space and will ignore PDGF. Conversely, the PDGF receptor will only react to PDGF and will never detect EGF or a dozen other kinds of growth factors that a cell may encounter.

A growth factor floating in the extracellular medium will bind directly to its receptor displayed on the cell surface. This binding will cause the receptor molecule to alter its overall structure; in response, the portion of the receptor that extends into the cell interior emits biochemical signals that persuade the cell to start growing. How does all this detail tell us something useful about cancerous growth?

SIGNAL-PROCESSING CIRCUITRY
....

The cell's decision to grow is the end product of long and complex deliberations. A nongrowing cell must receive and process a number of growth-stimulatory signals, notably those conveyed by growth factors, and assess whether their strength and number warrant entrance into an active proliferation phase. In addition, neighboring cells may release growth-inhibitory signals, which are also transmitted into the cell via specialized surface receptors. They too figure prominently in the final equation that determines whether or not the cell will undertake proliferation.

Decision-making like this demands a complex signal-processing apparatus inside the cell. A useful metaphor is an electronic circuit board constructed as a network of components that operate like relays, resistors, transistors, and capacitors. Each of these components is a logical device that receives signals from other components, processes and interprets these signals, and then passes them on to yet other components in the circuit.

A component of such a circuit may operate in a binary fashion. If it receives enough incoming signals, it will emit a signal to yet another component; if it fails to receive an adequate dose of such signals, it will remain silent. It will either be fully on or fully off. Alternatively, a signal-processing component may operate like an analog device: Greater fluxes of incoming signals will cause it to emit proportionately larger outgoing signals. Computers are assembled from such simple elements, which confer enormous signal-processing capacity when arrayed appropriately.

Inside the living cell, the signal-processing components are proteins rather than silicon diodes and capacitors. Like their high-tech counterparts, each of these proteins is endowed with complex signal-processing capabilities. In the jargon of the biochemists, these proteins are capable of "signal transduction," in that they receive signals, filter and amplify them, and then pass them on to other components.

Often these circuits form linear arrays that function much like molecular bucket brigades. A protein at the top of the brigade passes a signal to the next protein down the line, which in turn responds by transmitting the signal yet another step down. Biochemists call these chains of command "signal cascades." In a living cell, the proteins at the top of these cascades are the growth factor receptors. When activated by binding growth factors, these receptors energize a chain reaction that reaches deep into the cell, transmitting signals to the heart and mind of the cell— its nucleus.

One good example of a signal-transducer is the protein made by the normal *ras* proto-oncogene. It sits near the cell periphery, on the inner surface of the cell's membrane, patiently awaiting prompting from a nearby growth factor receptor. When the correct factor binds to this receptor, the latter then passes a signal through the plasma membrane into the cell interior or cytoplasm. In the cytoplasm, a part of the receptor emits a growth-stimulatory signal that is transmitted via intermediates to the *ras* protein. The *ras* protein in turn becomes activated and sends out signals to a protein one step further down in the signaling cascade. This downstream partner is a protein made by the *raf* proto-oncogene. The protein made by the normal *src* proto-oncogene acts in a similar way, a link in a long and complex signaling chain.

The presence of proto-oncogene proteins in the normal signaling cascade of the cell hints at the mechanisms that oncogene proteins use to create cancer. These proteins, in their normal or cancer-causing configurations, sit astride the major signal-processing pathways of the cell. In these strategic positions, they are well poised to influence the cell's behavior.

After growth-stimulating signals are passed through the cytoplasm, they arrive in the nucleus where they impinge upon the machinery that regulates the expression of genes. In particular, these signals encourage the readout of a number of genes, enabling the cell to make proteins that were, until that moment, absent or present only in low amounts. These new proteins become the servants of change in the cell; they scurry about, preparing the cell to move from quiescence into active growth.

Although cancer researchers played a role in the discovery of these signaling cascades, much of the information derived from other sources, in particular research on the genes controlling the growth of a single-cell organism—common baker's yeast—and still other genes controlling the development of the fruit fly eye and the vulva of a tiny earthworm. This theme appears repeatedly in the story of cancer research: Some of the biggest leaps forward have come from unexpected sources, from researchers whose work seemed unconnected with the problem of human cancer. In this instance, the uncovering of the human growth signaling cascade was aided by this cascade's very ancient lineage. It appears in very similar form in the cells of all animals and in clearly recognizable form in yeast cells.

Our ancestors and those of the fruit fly went their separate ways more than 600 million years ago. The common ancestor shared by us and baker's yeast probably lived more than a billion years ago. Once this signaling machinery evolved in an ancient progenitor, it became a fixed, unchangeable part of cells, vital to their continued existence, in particular their ability to regulate proliferation and differentiation.

Researchers have often taken advantage of this constancy: Rather than studying signaling inside human cells, which are often difficult to manipulate, they have turned to simpler organisms to learn basic truths about how life is organized on this planet. As described in the next chapter, this circuitry of ancient lineage malfunctions in cancer cells. Indeed, the changes in signal processing occurring in human cancer cells are only minor variations on a grand, billion-year-old theme.

CHAPTER 11

A BREAKDOWN OF ORDER:
SUBVERTING NORMAL
GROWTH CONTROL

The details of the cellular signaling circuitry have been pieced together over the past ten years. The plan of this circuitry provides the key to understanding the cell growth deregulation that creates human cancer. It also allows us to relate growth deregulation to the actions of specific genes. Proto-oncogenes and oncogenes form the blueprint of this circuitry, specifying its component parts—the signal-transducing proteins. When the genetic blueprint is intact, the circuitry operates flawlessly, enabling the cell to make decisions about growth and quiescence that are invariably correct. But when the blueprint is damaged through mutation, then certain components of the circuitry malfunction and disrupt the entire decision-making process. Cancer is a disease of faulty information processing deep inside cells.

We have already discussed one consequence of disrupting this signal-processing circuit: The growth of cancer cells is liberated from its normal dependence on external growth-stimulating factors. This liberation is achieved by oncoproteins through a simple trick. They activate the signal-processing circuitry by mimicking the signals that arise after a normal cell encounters growth factors. In effect, oncoproteins fool the cell into thinking it has encountered growth factor molecules.

Oncoproteins do this in several ways. One type of onco-protein induces the cancer cell to release growth factors into its immediate surroundings. This would seem to be a point-less exercise, until one realizes that these growth factors may then turn back and stimulate the growth of the same cell that has just released them. By encouraging cells to make their own growth factors, oncogenes and their protein products free these cells from dependence on growth factors of exter-nal origin. In effect, such oncogenes transform cells by caus-ing them to stimulate their own growth continuously. Evidence of this strategy is provided by various human tu-mors that release substantial amounts of PDGF and EGF into their surroundings.

The genes that specify growth factor receptors may also play an important role in creating human cancers. A mal-functioning receptor may delude a cell into thinking that it is swimming in a sea of growth factors, when in fact none may be present. As before, this will drive incessant growth.

Receptors can malfunction in at least two ways. The proto-oncogenes encoding growth factor receptors may suffer mutations that cause the receptor molecules to assume new shapes and structure. Such malformed receptor molecules may release a steady flood of growth-stimulating signals into the cell even when they have not found any growth factor. Some breast cancer cells, for example, make a shortened EGF receptor that fires continuously even in the absence of EGF.

Some human cancer cells display excessive numbers of receptor molecules. When receptor molecules are present in abnormally high concentrations on the cell surface, they may congregate and begin to fire spontaneously. This is ex-tremely effective in driving cell proliferation. For example, breast carcinoma cells that express abnormally high levels of the EGF receptor and its cousin receptor, often termed *erbB2/neu,* grow aggressively and are often unresponsive to therapy. The EGF receptor may also be overexpressed in glioblastomas (brain tumors) and stomach carcinomas. Here again, it may induce especially malignant growth.

Malfunctioning of the *ras* protein presents another way by which cell growth can be liberated from its usual dependence on external growth factors. As described above, the normal *ras* protein sits in the cell cytoplasm, awaiting a signal from a growth factor receptor. After receiving a pulse of signals from a receptor, the *ras* briefly enters an excited state, sending stimulatory signals further into the cell. Soon it shuts itself off and reverts to its resting state. This shutdown ensures that the downstream signaling machinery receives only a limited dose of growth-stimulatory signals.

The protein made by the *ras* oncogene differs subtly from its normal counterpart. Like the normal *ras* protein, the *ras* oncoprotein becomes excited by a growth factor receptor and responds by passing signals on to target proteins downstream in its signaling cascade. But unlike its normal counterpart, the oncogenic form of the protein lacks the ability to turn itself off! It remains in an activated state for an indefinite period, flooding the cell with an unrelenting stream of growth-stimulating signals.

The protein made by the normal *myc* gene lives in the cell nucleus, where it operates to induce the expression of other growth-promoting genes. In the absence of extracellular growth factors, a cell will make almost no *myc* protein. But within an hour after encountering growth factors, a cell will begin to churn out large amounts of *myc* protein, which will enable the cell to read out information in a number of genes that are vital to growth.

The *myc* oncogene behaves very differently from the normal proto-oncogene version. This oncogenic version of the *myc* gene is turned on constantly, often at high levels, driving incessant growth even when growth factors are absent.

Oncogenic versions of the *myc* gene have been found in a variety of human tumors. Some cancers achieve this constant, intense expression by increasing the number of copies of the *myc* gene. Instead of the two *myc* gene copies present in the normal cell, certain kinds of tumor cells may contain dozens. This amplification in the copy number of the *myc* gene seems to liberate *myc* from normal regulation, resulting in its high, constant expression. In other cancers, the *myc* gene becomes fused with a second gene; this second gene now imposes unnatural control on the expression of *myc*. In both instances, *myc* activity is liberated from its usual dependence on growth factor stimulation. The resulting intense production of the *myc* protein drives unrelenting cell growth.

A close cousin of the *myc* gene, termed N-*myc*, plays a big role in a childhood cancer. In cells of low-grade, relatively benign cases of childhood neuroblastoma—a tumor of the peripheral nervous system—the N-*myc* gene is represented in its usual two gene copies. In more advanced tumors, however, the number of N-*myc* gene copies may be increased to ten, twenty, or even a hundred per cell. These extra gene copies seem to be directly responsible for the increased aggressiveness of these tumors. Moreover, extra N-*myc* gene copies in the cells of a neuroblastoma are highly predictive of a poor response to therapy.

A BREAKDOWN IN COMMUNICATION: LOSS OF TUMOR SUPPRESSOR PROTEINS

Oncogene proteins stimulate the same signaling circuits that are normally activated by the cell in response to external growth factors. But unlike their normal counterparts, oncoproteins activate these circuits continuously and in the absence of any external growth-stimulating signals, driving unceasing cell proliferation.

But oncogene function is only one side of the coin. The tumor suppressor genes play an equally important part in creating human tumors. As discussed earlier, these genes and their encoded proteins, which act as brakes on cell proliferation, are lost during the multistep progressions that lead to human tumors. Such negative control implies mechanisms that are diametrically opposite from those of the oncogenes.

How do tumor suppressor proteins normally operate in the cell? At one level, their functioning, like that of the oncoproteins, can be described very simply. Cells receive two kinds of growth-regulating signals from their environments—those that stimulate growth and those that inhibit it. Cells must respond to the inhibitory signals using a signal-processing machinery that is as complex as that allowing them to respond to stimulatory signals. Many tumor suppressor proteins are components of the machinery that confers responsiveness to external growth-inhibitory signals. When tumor suppressor proteins are lost, the cell loses its ability to respond properly to these inhibitory signals and continues blithely on, proliferating even though its environment is telling it loudly and clearly to stop.

Once again, gene mutations disrupt the communication between a cell and its surroundings. In the case of tumor suppressor genes, the mutations deactivate or destroy gene function rather than enhancing it. Because tumor suppressor research is still in its infancy, we know little about how many suppressor proteins function, but certain facts already begin to stand out. Like oncogene proteins, the suppressor proteins operate at many sites from the cell surface all the way to the nucleus. What follows are several particularly interesting examples of tumor suppressor function.

At the cell surface lies a series of receptors that enable the cell to sense growth-inhibiting signals. The best studied of these signals are those carried by TGF-ß (tumor growth factor–ß). Like the growth-stimulating factors, TGF-ß is made from protein chains that are released by one cell, move through intercellular space, and impinge on a target cell, which may respond by shutting down its growth.

A variety of tumor cells seem to escape growth inhibition by TGF-ß. Unlike normal cells, these cancer cells seem oblivious to the presence of this factor; they continue growing under conditions where their growth is being strongly discouraged by TGF-ß.

Virtually all cells display specific receptor molecules on their surfaces that enable them to detect the presence of TGF-ß in the surrounding fluids. These TGF-ß receptors are structured much like the growth factor receptors. They extend one end into the extracellular space, thread through the surface membrane of the cell, and extend a second, signal-emitting structure into the cell interior.

Several types of cancer cells seem to have shed their normal complement of TGF-ß receptors. It is unclear how retinoblastoma cells, for example, lose these receptors, but this loss provides clear growth advantage to these cells. Normal retinal cells experience substantial levels of TGF-ß in the back of the eye. Lacking appropriate receptors, the retinoblastoma tumor cells are oblivious to TGF-ß and therefore ignore its orders to stop.

In individuals suffering from HNPCC, the precise mechanism of TGF-ß receptor loss is clear. In these people, a gene that serves as blueprint for one of the TGF-ß receptors suffers mutational damage. This gene falls victim to the faulty DNA repair machinery that afflicts cells of individuals suffering from HNPCC. Because of inadequate DNA repair, the DNA sequences of their TGF-ß receptor genes become scrambled, causing the encoded receptors to become nonfunctional.

Like retinoblastoma cells, these colon cancer cells become unresponsive to TGF-ß inhibition. Such escape from inhibitory signals would appear to be highly advantageous to tumor cells that find themselves in a Darwinian struggle for proliferative advantage.

The *NF*-1 tumor suppressor gene provides a quite different picture of how cells regulate their own growth. Individuals inheriting a defective *NF*-1 gene suffer from neurofibromatosis, which is manifested in numerous benign growths throughout the body, some of which may progress to malignancy. The *NF*-1 gene specifies a protein that operates in the signaling pathway through which the *ras* protein funnels its stimulatory signals. This would seem like a paradoxical place for a tumor suppressor protein. The paradox is resolved upon close examination of functioning of the *NF*-1 protein: It works to shut down the *ras* protein.

After the *ras* protein is stimulated into an excited state by a growth factor receptor, the *NF*-1 protein may ambush the *ras* protein and deactivate it before *ras* has had the opportunity to emit growth-stimulating signals. This preemptive attack on the signaling pathway damps down growth-stimulating signals in the cell. In the absence of the *NF*-1 protein, too many stimulatory signals flow into the cell's nucleus, again driving cell proliferation.

In the cell nucleus are found a series of other suppressor proteins including the proteins made by the *p16, Rb, p53* genes and *WT*-1 tumor suppressor genes. The first three of these act as brakes for the cell cycle clock apparatus, which will be discussed later; the *WT*-1 protein regulates the expression of cellular genes that remain to be identified. The precise mechanisms through which other tumor suppressor proteins operate are still unclear.

The picture painted for tumor suppressor proteins has as many colors and forms as the earlier descriptions of oncogene proteins. The suppressor proteins operate in various locations within the cell. They act through a variety of molecular mechanisms to shut down cell proliferation. But they are united by a common theme: The loss of each of these suppressor proteins renders the cell incapable of responding properly to growth-inhibitory signals. Cell proliferation continues when, by all rights, it should cease.

It might be tempting to depict the proto-oncogene and tumor suppressor proteins as forming two distinct, parallel signaling circuits in the normal cell, one dedicated to growth promotion, the other to its suppression. This view is mistaken. In truth, the two kinds of proteins form a single common circuitry having both positively and negatively acting components. Within this circuitry, the two types of proteins counterbalance each other, making possible the finely tuned decisions that are essential for the cell to participate in the construction and maintenance of normal tissue architecture.

....

CHAPTER 12

IMMORTALITY: AVOIDANCE OF ALMOST CERTAIN DEATH

Runaway oncogenes and defective tumor suppressor genes would seem to provide a full explanation of why cancer cells grow uncontrollably. Mutant versions of these two classes of genes conspire to make cells grow when, by all rights, they should be resting and inactive. This collaboration was illustrated graphically by the case of colon carcinoma, which often involves mutations of the *ras* oncogene and three tumor suppressor genes. Still, this view fails to address an important reality of cell biology: Tissues have two very different ways of limiting cell proliferation. The first strategy depends on depriving cells of growth factors or exposing them to signals that discourage growth; these conditions result in nongrowing, quiescent cells. This strategy, so essential to the maintenance of normal order within a tissue, is disrupted by changes in various proto-oncogenes and tumor suppressor genes.

The other strategy for constraining cells involves a much more drastic measure: Tissues hold down cell numbers by inducing them to commit suicide. Elimination of these sacrificial lambs represents an equally important way of controlling the size of cell populations.

Cells in many tissues throughout the body may be consigned to death for a variety of reasons. One condition that will provoke cell death is illustrated by a simple experiment. If cells are taken from one or another tissue and cultured in dishes, they will double a limited number of times before they stop growing, become sickly, and eventually die, steps that are termed cell senescence and crisis. Human cell populations, as an example, will often double once a day for fifty to sixty days before they stop growing. This barrier to unlimited proliferation is often termed "cell mortality."

Cell mortality appears to be an important anticancer defense mechanism. Normal tissues seem to endow their cells with a limited number of doublings in order to erect a barrier to tumor development. This barrier ensures that incipient tumor cell populations will double only a limited number of times before they exhaust their allotment of doublings and cease growth.

The cell mortality barrier must be breached by developing tumor cell populations. A population of premalignant cells that lacks this ability to divide endlessly will be unable to expand into a mass of substantial, life-threatening size. Indeed, when tumor cells are placed in culture, they show an ability to multiply indefinitely, an indication that these cells have become "immortalized."

Until recently, this phenomenon of cell mortality represented a major mystery for biologists. How do cells know when to stop growing and become senescent? How does a cell lineage know when it has exhausted its allotment of doublings? Cells seem to carry some record or collective memory of their past history. Some counting device registers each time the cells in a lineage pass through a cycle of growth and division, tallying the number of cell generations that separates cells in a tissue from their distant ancestors in the early embryo.

There are other examples for such generational counting. In some Chinese families, the first of the given names of the children signifies the generation number that separates them from their earliest recorded ancestor in the family chronicle. The cells in our tissues must also carry markings like these names, which tell them about their place in the history of that organism, beginning with the moment of conception. These markings must be registered by a "generational clock." When this clock reaches a predetermined limit, having tallied out a certain number of generations, it sounds an alarm which tells the cell to stop growing and become senescent. In some fashion, cancer cells learn to ignore this alarm and continue their endless cycles of growth and division.

The counting mechanism used by the generational clock has long defied description. Recent exciting research carried out in a number of laboratories has finally revealed the molecular basis of the generational clock. It represents a most clever and unexpected solution to the problem of counting cell generations.

These discoveries about generational counting, like many others described here, came from research areas that seemed to have little or nothing to do with human cancer. They began with observations made in the 1930s by two geneticists, Barbara McClintock and Hermann Muller. These two concluded that the chromosomes of corn and fruit flies contain specialized ends that protect the chromosomes from fusion with one another and from fragmentation. Muller called these ends telomeres. Telomeres function much like the shields at the ends of shoelaces that prevent fraying. Every human chromosome is a linear structure and therefore has two telomeric ends.

Almost forty years later, in 1972, James Watson, the co-discoverer of the DNA double helix, added a critical piece to this story. By then, the mechanisms of cell division including the process of DNA replication were understood in some detail. Each time a cell prepares to divide, it copies its DNA to guarantee that each of its daughters will be provided with an equal dowry of genetic information. Earlier, we described how precise DNA copying and editing can be, with accumulated errors of less than one part in a million. But Watson noted one glaring exception to this rule of efficient and faithful genome replication: Because of the biochemical mechanism used by the DNA polymerases—the enzymes responsible for DNA copying—the extreme ends of the chromosomal DNA are never copied properly. The result is that the telomeres, which form these ends, shorten by one hundred or so bases each time a cell copies its DNA.

Several years later, Elizabeth Blackburn, a geneticist working on paramecia, single-cell pond protozoans, discovered the structure of telomeres. They are made of DNA double helices like the rest of the chromosome but have a very unusual sequence composition, being assembled from many copies of the same DNA sequence repeated over and over again. In human chromosomes, the telomeres are composed of the base sequence TTAGGC repeated a thousand times or more.

These observations, taken together, led to a major puzzle: How do protozoans like paramecia double indefinitely year after year if their DNA copying apparatus is incapable of regenerating the essential telomeric ends of their chromosomes? By 1984, Blackburn's group had the answer. Paramecium cells express a specialized enzyme, termed telomerase, whose function is to regenerate the ends of telomeres by adding DNA sequence repeats, thereby compensating for the inability of the general DNA copying machinery to do so.

In the 1970s, unbeknownst to these researchers, the Soviet geneticist A. M. Olovnikov had proposed a theory linking telomeres to the phenomenon of cell mortality. He proposed that normal cells in the mammalian body, unlike paramecium cells, are unable to regenerate their telomeres. For that reason, after thirty, forty, or fifty cell doublings, the telomeres are worn down, shortened so much that they lose their ability to protect the vital tips of the cells' chromosomes. The chromosomes, as a consequence, fuse end-to-end with one another, creating the genetic disarray that causes cells to stop growing and eventually die. It is this telomeric collapse that sounds the alarm telling a cell it has exhausted its allotment of doublings.

Olovnikov's speculation was eventually validated. By the early 1990s, work in a number of laboratories revealed that the telomeres of human cells shorten progressively as these cells pass through repeated cycles of growth and division. Eventually, lacking adequate telomeres, these cells enter first into senescence and then crisis, the latter process resulting in their death.

Not all the cells in the body are fated to suffer telomeric collapse and resulting chromosomal fusion. At least one lineage in the body is spared this fate and is therefore guaranteed immortality—the germ cells, sperm and eggs. Germ cells must be able to perpetuate themselves without limit in order to ensure the genetic immortality that allows genes to be passed in repeated cycles from one organism to its offspring and thence to succeeding generations. Such transmission without time or generational limit is necessary for the continuity of a species over millions of years.

How do germ cells ward off the crisis triggered by telomeric collapse? Unlike almost all other cells in the body, they express the telomerase enzyme, which compensates for any shortfalls created by the DNA polymerase. It seems that the telomerase enzyme is expressed by many, perhaps all cells in the early embryo shortly after the egg is fertilized. Soon, however, production of this enzyme is extinguished in the cell lineages leading to most tissues, the germ lineage being the exception. This extinction imposes a limit on the potential of these many lineages to proliferate—a powerful barrier to cancer development.

Cancer cells violate this nicely crafted scheme by resurrecting the telomerase. All cells in the body, normal and malignant, carry the genetic information for making telomerase, but this information is suppressed during embryonic development in most normal cell lineages. By unknown means, cancer cells acquire access to this hidden information in their DNA and use it to make telomerase once again.

The telomerase gene is the apple from the Tree of Knowledge, forbidden to most normal cells in the body. Once cancer cells have accessed the gene and resurrected the telomerase enzyme, they are able to regenerate and maintain the ends of their chromosomes indefinitely, guaranteeing them unlimited replicative potential. Now their ability to multiply is only limited by one remaining barrier—the ability of the cancer patient's body to sustain their ever-growing numbers.

In some tumors, telomerase appears relatively late in the multistep progression of normal cells into cancer cells—only after an evolving population of premalignant cells begins to exhaust its allotment of generational doublings. The appearance of telomerase in these cells depends on the expression of a gene specifying a key component of this enzyme. Ongoing research is focused on determining how this gene is shut down in normal cells and how its expression becomes possible in tumor cells.

Earlier, we saw how activation of oncogenes and deactivation of tumor suppressor genes profoundly change the foreign relations of a cell—its interactions with its surroundings. Telomerase resurrection represents an altogether different kind of change, strictly a domestic housekeeping matter, the result of a cell tinkering with and overcoming its own built-in limitations.

The cloning of the telomerase gene, which occurred in 1997, has excited those interested in developing new kinds of anticancer therapies. Attempts at creating effective anticancer drugs have been thwarted repeatedly by the substantial similarities between normal cells and cancer cells. While we have described a number of genetic differences that distinguish normal and cancer cells, these mutations affect only a minute fraction (less than 0.01 percent) of the genome. The vast majority of genes are identical in normal and cancer cells. Similar genetic repertoires dictate similar overall appearance, behavior, and biochemical makeup.

These similarities explain why almost all drugs that have been used experimentally to kill cancer cells have had devastating effects on their normal cousins. These drugs have lacked selectivity—the ability to target cancer cells while leaving normal cells relatively unscathed. Very few anti-tumor drugs under development for use in the clinic survive the preliminary tests designed to uncover any toxic effects they may have on normal tissue.

But telomerase presents a rare exception to this general rule of commonly shared traits, and therefore may represent an Achilles heel of cancer cells. It is present in these cells and essential for their growth; most normal adult cells lack the enzyme and therefore must not depend on telomerase for their continued existence. This suggests an obvious strategy for drug development: the creation of drugs that attack and inhibit telomerase while leaving the thousands of other enzymes in cells untouched. Such a highly targeted drug may stop cancer cells in their tracks while having little if any effect on normal cells.

There is at least one fly in this ointment. Some kinds of normal cells, notably white blood cells, have been found to express telomerase under certain conditions. This suggests that these cells may require telomerase for their growth, raising the possibility that anti-telomerase drugs may affect certain types of normal cells and exhibit undesirable side effects. Nonetheless, on balance, the development of such drugs remains attractive. It will take another decade before we will know whether antitelomerase drugs can be made and, if so, whether they will prove useful for treating tumors.

CHAPTER 13

ASSISTED SUICIDE:
APOPTOSIS AND
THE DEATH PROGRAM

The generational clock represents one way by which the body holds down the number of its cells. But there is at least one other equally powerful strategy used by the body to limit cell multiplication. Our tissues can induce superfluous or defective cells to kill themselves. Cancer cells must learn to evade this death machine. It represents another trick used by the body to foil cells on their way to becoming highly malignant. The ability of an organism to eliminate selected cells in its tissues has long been apparent to embryologists. Perhaps the most dramatic example comes during the development of our hands. Early on, extensive tissue webs hold the fingers together. Later, most of the cells in these webs die, leaving only the small vestigial webs at the bases of our fingers. But there are many other less visible sites in the developing embryo where cells are eliminated as well, often in large numbers. In the brain, for example, large numbers of embryonic nerve cells that fail to form proper connections are sacrificed.

This practice of eliminating unwanted cells is of ancient vintage, being apparent in a primitive animal that may, in many respects, resemble our ancestors of 600 million years ago. In the tiny worm *Caenorhabditis elegans*, a total of 1,090 cells are formed by repeated rounds of cell division following fertilization of the egg. Of these, exactly 131 die in predictable sites in the embryo and at precisely scheduled times of embryonic development.

Most biologists assumed until recently that these cells died through slow disintegration, victims of attrition, starvation, or damage to some of their vital parts. Such slow death was thought to resemble the necrosis provoked by certain poisons. During the course of necrosis, cells swell, their internal parts disintegrate, and the cells burst.

We now know that many cells take a very different route to the grave. They actively kill themselves in a quick, certain, and highly stereotypical fashion. Their suicide depends on a built-in death program. In 1972, Andrew Wyllie, one of the codiscoverers of programmed cell death, named it "apoptosis." The term refers to a Greek word that describes the shedding of leaves by a tree. Once the death program is triggered, a cell will die and disintegrate, and the cell fragments will disappear in less than an hour.

The apoptotic death program seems to be wired into the control circuitry of every human cell, a self-destruct mechanism much like the explosive devices rocket manufacturers build into satellite launchers. An errant trajectory will cause the ground controller to trigger destruction of the rocket. Similarly, an errant or unwanted cell will be targeted for destruction, the result of a decision made by the surrounding tissue or by the cell's own internal control machinery.

A cell going through apoptotic death is not a pretty sight. First, its nucleus shrinks. Then its outer membrane begins to herniate at many points. Soon its chromosomal DNA is chewed into small pieces. Finally, the cell explodes into small fragments that are quickly gobbled up by neighbors. No trace of the cell is left behind. It is as if it had never existed.

The frequent loss of cells by apoptosis during embryonic development seems counter to intuition, which portrays embryo development as a period of vibrant expansion. It would seem that a well-designed embryo should create only the cells that it requires for the formation of its component tissues, no more and no less. But in truth, embryonic development is profligate and inefficient. In many locations in the developing embryo, far more cells are produced through cell division than will ever be used in the final developed organ or tissue. Some of these cells create tissues that are evolutionary vestiges having no use in the modern organism. Others are the results of failed attempts during development at constructing proper tissue architecture. Apoptosis is used much like the sculptor's chisel to remove unwanted material.

Recent research has made it clear that the body continues to employ apoptosis throughout life, not just during embryonic development. In the immune system, large numbers of cells that fail to develop the capacity to manufacture proper antibodies are discarded. Many adult tissues maintain their architecture through constant winnowing by apoptosis.

Mammalian cells exploit the apoptotic death program in other situations as well. Cells infected by a variety of viruses attempt to activate their apoptotic program. Their motive here is clear: By sacrificing themselves quickly, these cells will deprive the virus of a suitable host in which to multiply, thereby aborting the viral growth cycle. This altruism spares nearby cells from subsequent rounds of infection. To neutralize this defense mechanism, many viruses have evolved countermeasures that rapidly block their host cell's apoptotic response.

Apoptosis is also the fate of overtly defective cells in the body, especially those that have sustained heavy, irreparable damage to their DNA. In some fashion, still unclear, cells are able to sense when they have sustained serious lesions in their genomes. Rather than attempting to repair the damage, cells are hardwired to commit suicide.

Many cells that suffer only minimal damage and remain viable are nonetheless consigned to apoptotic death. At first glance, these suicides appear wasteful. Tissues dissipate valuable resources by continually creating new cells to replace those that, though only minimally defective, have been weeded out nonetheless. But in the end, this spending of resources represents a far smaller concern than the danger posed by the continued presence of a damaged, possibly mutant cell. This begins to suggest an important role of apoptosis in preventing cancer by rapidly eliminating errant cells scattered throughout the body's tissues.

Subtle irregularities in the internal growth-regulating circuitry of cells can trigger the death program. Such irregularities may occur within cancer cells, associated with metabolic imbalances and inappropriate growth signaling. For example, the introduction of a *myc* oncogene into a normal cell seems to cause a signaling imbalance that provokes many cells to trigger their apoptotic death program. This suggests that many cells that have acquired a *myc* oncogene through some accidental mutation may be eliminated rapidly through apoptosis. A small minority of these cells may, through one strategy or another, evade the almost inevitable suicide program. Indeed, all cells may be hardwired to kill themselves in the event that an oncogene becomes activated within them. In effect, the body has set up trip wires in all its cells. These alarm devices constitute barriers to tumor formation by consigning incipient cancer cells to quick and certain death.

We conclude that a cell en route to becoming cancerous must carefully negotiate the minefield of apoptosis. Having acquired a growth-promoting oncogene, it must somehow avoid apoptotic death. That avoidance may sometimes be achieved by a second gene mutation. For example, the apoptosis that is usually triggered by an activated *myc* oncogene may be prevented in certain circumstances by the subsequent activation of a *ras* oncogene.

The role of mutation in warding off apoptotic death is most apparent in the immune system. As mentioned earlier, immune cells are eliminated by apoptosis if they fail to develop the ability to make proper antibodies. Up to 95 percent of certain classes of lymphocytes, key players in the development of the immune system, appear to be discarded in this fashion. Here we witness a tissue eliminating cells that are not overtly defective and threatening but merely unproductive.

Subversion of this death program in lymphocytes can also lead to cancer. Lymphocytes can acquire the ability to escape apoptosis by activating the *Bcl*-2 oncogene, which is specialized to prevent the triggering of the death program. Populations of lymphocytes carrying an activated *Bcl*-2 oncogene will begin to expand substantially, having escaped their almost inevitable apoptotic fate. These cells are not malignant; they simply accumulate in uncommonly large numbers. However, some of these excess cells may, years later, sustain other mutations, including those that activate the *myc* oncogene, and then develop truly malignant progeny that result in lymphomas. Accumulating evidence suggests that yet other kinds of cancer cells ensure their own long-term survival by activating the *Bcl*-2 oncogene, either by mutating this gene or by causing it to be expressed excessively.

In all types of human cancer, simple arithmetic determines whether or not premalignant cells will eventually succeed in creating a full-blown tumor. To form life-threatening tumors, cells must increase their ability to proliferate and, at the same time, find means of avoiding death. Some groups of premalignant cells may well succeed in increasing their rate of multiplication by acquiring an activated oncogene but may fail to address the threats of apoptosis and senescence;

any gains they make through increased proliferation may be neutralized by equal or greater rates of cell death. The net result will be a cell population of constant or even diminishing size. Only when the problem of cell death is solved can such a cell population begin to expand rapidly, generating a Malthusian explosion.

PP53 AS GUARDIAN OF THE GENOME AND MASTER OF THE DEATH PROGRAM

A cell's decision to commit apoptotic death or to avoid it is influenced by a number of central controllers, most notably the *p53* tumor suppressor gene. This gene, acting through its protein, serves as the arbiter between life and death, an ever-watchful guardian that monitors the cell's well-being and sounds the death alarm if the machinery of the cell is damaged or begins to run amok. Nowhere is the role of *p53* more apparent than in the cellular responses that occur following damage to the cell's DNA.

We have described how the genome of the human cell is constantly under attack, pelted by a hailstorm of damaging chemicals and miscopied by wayward DNA polymerases. Cells can respond to the resulting genetic damage in two ways: They may try to repair it using the repair apparatus we described earlier, or they may throw in the towel and enter into programmed cell death. If the mutational damage is minor, the cell will try to fix it; if it is massive and exceeds the capabilities of the repair apparatus, the cell chooses its only remaining option—apoptosis.

Normally, cells rely on the *p53* protein to help them respond to DNA damage. Like other tumor suppressor proteins, it operates like a brake shoe. The levels of the *p53* protein inside the cell build up rapidly within minutes after DNA damage is sensed. Once accumulated to high concentration, the *p53* protein acts as an emergency brake, rapidly shutting down the cell's progress through its growth cycle.

In cases of minor DNA damage, this halt may be temporary, a short respite from the rapid race around the cell growth cycle. While the *p53* protein blocks the cell's advance, the repair apparatus has time to search out and repair damaged base sequences. Once the damage has been erased, *p53* will back off, permitting the cell to begin growing once again.

The logic behind this response is simple. The temporary halt prevents the cell from entering into the phase of its growth cycle when its DNA is scheduled to be copied. The *p53* protein will permit entrance into this DNA copying phase only after DNA damage has been successfully repaired. This ensures that the replicating enzymes—the DNA polymerases—do not inadvertently copy damaged DNA, thereby generating equally defective copies that will be perpetuated as mutations passed on to later cell generations.

Should DNA damage be massive, the response will be quite different. The *p53* protein will, as before, accumulate in large amounts in the cell. Once again, the cell will be stopped in its tracks. However, the cell's damage assessment program will now gauge the extent of genetic devastation and decide to activate an additional response: The apoptosis program will be engaged. The outcome is sure and swift: The cell will die within an hour or so, and with it will die all its newly damaged genes. Sacrificing this cell through apoptosis represents a substantial expense in terms of biochemical resources, but in the long run this option is far less costly to a tissue than the presence of a mutant, potentially malignant cell in its midst.

For the incipient cancer cell, inactivating the *p53* gene through mutation has obvious benefits. Once a cell has knocked out its *p53* gene, this damage response pathway is crippled. As a consequence, this cell and its descendants can continue to proliferate even after sustaining substantial damage to their genomes. Lacking functional *p53,* these cells will race ahead and copy their still-damaged DNA, incorporating unrepaired lesions into the newly made genome replicas. Resulting mutant genomes will then be passed on to descendant cells.

Without *p53* on duty, the usually slow mutational processes that activate proto-oncogenes and deactivate tumor suppressor genes will be accelerated substantially. Since these mutations are the events that limit the rate of tumor progression, the evolution of a tumor cell population will be greatly accelerated and the time of appearance of a full-fledged malignancy will be advanced. In short, loss of *p53* may be as devastating for genomic stability as a major defect in the DNA repair apparatus.

Normal cells grown in culture have a low, almost undetectable tendency to accumulate excess copies of their genes. However, in the absence of functional *p53,* their tendency to accumulate excess copies of genes is enhanced more than a thousandfold. Such gene "amplification," as described earlier, can result in increased copies of growth-promoting oncogenes such as *myc, erb* B and *erb* B2/*neu.* Amplification of these genes occurs frequently during the formation of a variety of cancers, including brain, stomach, breast, and ovarian cancers and childhood neuroblastomas.

Almost all kinds of tumor cells achieve immortalization, and *p53* inactivation aids in this process as well. Recall that the barrier to immortalization is telomere shrinkage and collapse. Once telomeres have shrunk to a small size, an initial alarm is sounded that tells cells to stop growing and enter into a senescent, nongrowing state. The shortened telomeres seem to be perceived by the cell as damaged DNA. *p53* appears to be mobilized in response to this genetic emergency and responds, as usual, by shutting down cell growth. Such cells will remain in this static state for long periods of time and are said to be senescent.

Cells lacking *p53* are able to continue growing in spite of this telomere shrinkage. They forge ahead, put senescence behind them, and multiply for another ten or twenty cell generations. By then, a second alarm is sounded, triggered by telomeres that have continued to shorten during this period and ultimately reach a critically short size. Such cells now die in large numbers. Only the rare variants that resurrect telomerase can escape this crisis, repair their telomeres, and become immortalized. While *p53* deactivation does not itself create immortalized cells, it enables tumor cells to reach the point where they can try for the golden ring—eternal life through telomerase resurrection!

Very recently, another aspect of *p53* deactivation has come to light. Cancer cells in the middle of a tumor mass often have poor access to the blood supply and thus to oxygen. They become anoxic—starved for oxygen—and cease growth. Most normal cells enter into apoptosis if they experience anoxia for an extended period of time. *p53* seems to mediate this response. However, if the *p53* gene is knocked out by mutation, as happens in many kinds of tumor cells, these cells can survive for extended periods, waiting for the time when they succeed in acquiring an adequate blood supply and oxygen and can resume rapid, unimpeded proliferation.

The state of a cell's *p53* protein also has direct implications for cancer therapy. Almost all the agents used to treat cancer—chemotherapeutics and radiation—operate through their ability to damage the DNA of tumor cells. The chemotherapeutics may react directly with the DNA bases, changing their structure, or they may interfere with the enzymes responsible for copying the DNA. X-rays often inflict irreparable damage on the double helix.

For three decades, the assumption was that these anti-cancer agents kill cancer cells by inflicting massive, widespread DNA damage. This damage, so the thinking went, would overwhelm the cancer cells' repair apparatus. These cancer cells would then stop growing and die because their chromosomal DNA had been shredded to pieces.

We now realize that anticancer therapies usually operate very differently. The doses of chemotherapeutics and X-rays that succeed in killing cancer cells do not in fact inflict massive destruction on their genomes. Instead, these treatments create just enough damage to activate *p53* and in turn the programmed cell death response. So cancer therapy does not succeed by bludgeoning cancer cells with a sledgehammer. Instead, it tweaks the cancer cells' control machinery, nudging these cells over the line that separates normal growth from apoptotic death.

This explains why *p53* is often a central player in determining the responsiveness of cells to anticancer therapy. As has been observed recently, cancer cells that have lost *p53* function are often more resistant to treatment, apparently because they are not readily coaxed into committing suicide. This may have substantial implications in the cancer clinic, where treatment strategies may soon be fine-tuned by knowing the status of the *p53* gene present in a patient's tumor cells.

....

CHAPTER 14

A CLOCK WITHOUT HANDS:
THE CELL CYCLE CLOCK APPARATUS

Every cell needs a well-functioning brain, in effect an executive who sits in some central office, receives information from branch offices, weighs all options, and makes the tough decisions. In fact, the range of possible decisions made by the executive brain of a cell is limited—whether the cell should grow, or differentiate into a specialized cell type, or die. If these critical decisions are made improperly by individual cells, the well-organized communities of cells that form human tissues will disintegrate into unruly throngs, with each member cell veering off on its own capricious trajectory. While the output of this executive decision-making machinery is limited, the information that informs these decisions is highly complex, originating from dozens of sources. These include the external signals brought to the cell by growth factors, chemical interchanges with neighboring cells, and physical contacts with adjacent cells and with the proteinaceous matrix that surrounds many cells. In addition, there is a wealth of internal housekeeping information, including periodic status reports on the health of the cell's DNA and the activity of the cell's metabolic machinery.

Somehow, this complex mix of information must be distilled, factored, and processed. Somehow it must converge on a single final decision, one that can only be made by a single, all-powerful decision maker. Over the past decade, the identity of this mysterious executive has been revealed. It is the cell cycle clock that operates deep within the cell's nucleus. This clock sits behind the executive desk, receives the complex information inputs, makes the hard decisions, and issues the orders.

The clock apparatus orchestrates the life of the cell—its cycles of growth and division. The active growth cycle of a cell can be divided into four discrete phases. A cell will often spend six to eight hours copying its DNA (the S phase) and three to four hours preparing for cell division (the G2 phase). Then begins the complex choreography of cell division, known as mitosis (the M phase), which takes only an hour.

After division, the two newly formed daughter cells will take ten to twelve hours to prepare for the next round of DNA copying during the G1 phase. Alternatively, cells in G1 may choose to exit the active growth cycle entirely and enter into a quiescent, non-growing state (the G_0 phase) in which they may remain for days, weeks, months, even years. This Rip van Winkle sleep is reversible. When provided with the proper signals, such sleeping cells will wake up and jump back into the active growth cycle. Within hours, they will be found racing around the track once again.

Human cells in active growth may move through this cycle once each day, though some race through it much more rapidly. By regulating the forward motion of the cell around this life cycle (usually called the cell cycle), the cell cycle clock determines the fate of the cell.

Not surprisingly, the operations of the cell cycle clock are disrupted in cancer cells. In these cells, the clock makes decisions that are, by normal standards, highly inappropriate. Instead of proceeding cautiously, carefully weighing the alternatives of growth and quiescence, the cell cycle clock will decide rashly and choose the growth option. In effect, the clock apparatus spins uncontrollably. Since it acts as the master controller of the cell, the cell around it will respond by growing and dividing without limit.

The central role of the clock is underscored by its position in the cell's signaling circuitry. Sooner or later, all of the signals received and processed by proto-oncogenes and tumor suppressor genes converge on the cell cycle clock. In effect, all the wires from the peripheral circuitry are strung into the nucleus and plugged into the clock. To understand the clock is to understand growth, normal and malignant.

The metaphor of a clock implies gear wheels and ratchets, but of course, inside a cell all complex machinery is assembled from proteins. In this instance, two types of protein components are involved, cyclins and cyclin-dependent kinases (CDKs). The CDKs, like all kinases, operate by attaching phosphate groups to target proteins. This phosphorylation of target proteins changes their functions, pushing them into a state of higher activity or switching them off. Because a kinase enzyme may paste phosphates onto many different target proteins, it can change many different processes in the cell simultaneously, in effect broadcasting a wide and powerful signal throughout the cell.

The CDKs that form the core components of the cell cycle clock are controlled by partner proteins that bind to them and direct them to appropriate targets. These partner proteins are the cyclins—the guide dogs of the CDKs. A CDK that lacks a partner cyclin will be blind and unable to phosphorylate any of its usual targets.

Cyclins come and go as the cell passes through the different phases of its growth cycle. When present, some cyclins will direct their partner CDKs to phosphorylate target proteins that are essential to the cell's ability to copy DNA; other cyclins may direct the phosphorylation of target proteins that cause the cell to divide. In the absence of these cyclin-

CDK partnerships, most of the cell's business shuts down and the cell retreats into a deep freeze—its G_0 state.

The subtlety in the operations of the cell cycle clock comes not from the cyclins and their partner CDKs. These two modular components are the clock's unthinking gears. It is the controllers governing them that provide the subtlety. By encouraging or inhibiting cyclin-CDK action, these controllers determine the motion of the clock.

In a normal cell, almost all of the important decisions that determine whether it will grow or not are made in the G1 phase of its growth cycle in the hours after cell division and before the next DNA copying begins. During this time window, the cell will commit itself to continuing growth, exiting the active growth cycle, or entering into a differentiated state where it takes on a new, specialized garb and simultaneously gives up all options of ever dividing again. Most of the cells in our body are in such "postmitotic" differentiated states. They go about their specialized jobs but will never grow and divide. Sadly, this describes the fate of the nerve cells in our brain, which die by the millions every day of our adult lives but are not replaced because their surviving neighbors have lost the ability to proliferate ever again.

Several well-studied tumor suppressor proteins operate in the middle of the cell cycle clock apparatus, acting as the subtle governors that brake various parts of its operations. The retinoblastoma protein, for example, acts as a master brake in the mid- to late part of the G1 phase. It erects an absolute block to further advance unless it is phosphorylated by the proper combination of cyclins and CDKs. Without such phosphorylation, the cell is stopped in G1 and is forced to exit from the active growth cycle. Tumor cells that lack the retinoblastoma protein proceed into DNA copying (the S phase) without a careful weighing of the factors that cause a normal, well-behaved cell to pause and consider its course of action.

The *p53* tumor suppressor protein has a hand in this machinery, too. As described earlier, *p53* levels increase in response to DNA damage. Once mobilized, *p53* causes a second protein, *p21,* to be made; the latter then shuts down the cell cycle clock apparatus by jamming itself into all the cyclin-CDK partner complexes.

Two tumor suppressor proteins, *p15* and *p16,* also act to control the clock. These two are almost identical twins; either one can block one of the critical CDKs that operate in the middle of the G1 phase of the growth cycle. They thus prevent the cell from advancing beyond the middle of G1. TGF-ß, the potent growth-inhibitory protein, exerts many of its effects through *p15.* Recall that TGF-ß binds to receptors at the cell surface. Once they have bound TGF-ß, these receptors send signals into the cell that can cause a thirtyfold increase in the *p15* braking protein; *p15* then shuts down the clock by blocking a critical CDK.

Individuals afflicted with familial melanoma often inherit defective versions of the *p16* gene. Lacking the ability to shut down the cell cycle clock under certain circumstances, their cells will continue to grow inappropriately. Very recent research indicates that loss or deactivation of the *p16* gene also plays a role in a wide variety of other cancers. Indeed, some laboratories report that *p16* activity is absent in more than half of all human tumors.

All of the signals that promote cell proliferation must eventually converge on the cell cycle clock. For example, the signals triggered by growth factors at the cell surface are funneled through the cytoplasm into the nucleus where they affect the operations of the clock machinery. Most important, growth-stimulating signals induce the production of high levels of cyclin D, one of the critical G1 clock components. Cyclin D, acting with a partner CDK, proceeds to phosphorylate and deactivate the retinoblastoma brake protein, which in turn enables the cell to advance into the next phase of its growth cycle.

Some time during the next decade, we will understand in exquisite detail how all of the signals that impinge on the cell surface affect the components of the clock apparatus, and how the clock, for its part, processes these conflicting signals, reaches decisions, and sends its marching orders out to the cell.

TICKLING THE CLOCK: THE GROWTH
STRATEGIES OF DNA TUMOR VIRUSES

Early in our story, we described a class of tumor viruses that are capable of infecting normal cells and transforming their newfound hosts into cancer cells. These retroviruses can induce cancer because they carry stolen cellular genes that they have remodeled into potent oncogenes. RSV provides the most notorious example of these cancer-causing agents. Recall that an ancestor of RSV invaded a chicken cell, kidnapped the cell's *src* proto-oncogene, and rapidly remodeled the captured gene into a powerful instrument for creating cancer. The study of the *src* oncogene led to the discovery of proto-oncogenes and in turn triggered the revolution in our understanding of cancer's origins.

Other kinds of tumor viruses succeed in creating cancer through a radically different strategy. These others have spent millions of years patiently crafting their own oncogenes. The quick-fix artists such as RSV carry their genes around in the form of RNA molecules. The patient artisans use DNA as their genetic material.

DNA and RNA molecules are equally able to encode genetic information. They are almost identically structured, and both are assembled from long runs of bases strung together end to end. DNA happens to be the molecule chosen by cells to store genetic information because it is so stable. But short-lived viruses are less constrained by considerations of long-term storage of genetic information, and for that reason some use DNA and some use RNA as their genetic data repositories.

This dichotomy creates two major kingdoms of viruses. The two kingdoms have evolved quite separately from one another. Their ways of parasitizing infected cells are very different. Also, RNA and DNA tumor viruses use totally different strategies for transforming infected normal cells into tumor cells.

The central agenda of the DNA tumor viruses, like that of all viruses, is simple and highly focused: They only want to make more copies of themselves. Although an occasional cell infected by one of these viruses may happen to become cancerous, causing cancer is only an accidental by-product of their drive to be fruitful and multiply.

In order to succeed in their agenda, the DNA tumor viruses must invade cells and parasitize their hosts' DNA copying apparatus, subverting and redirecting it to make replicas of viral DNA rather than cellular DNA. By using the DNA copying machinery of their host cells, these viruses spare themselves the trouble and expense of assembling their own.

But this parasitic strategy creates a quandary for these viruses, since most infectable cells in the body at a given time are in a resting (G_0) state outside of their active growth cycle. Such resting cells shut down much of their growth machinery, including the apparatus for copying DNA, and thus make poor hosts for an infecting DNA tumor virus. The challenge for these viruses is to persuade their newfound hosts to become more hospitable.

DNA tumor viruses solve this problem through sheer cleverness. After one of them invades a host cell, it will induce its host to leave the resting state and move into the active growth cycle. The awakened host cell now mobilizes its growth machinery, including its previously inactive DNA copying apparatus, anticipating use in the growth cycle. The virus, however, has other plans: It preempts this normal use by exploiting the host cell's DNA copying machinery to replicate its own DNA. The copied viral DNA is then packaged into new progeny virus particles that leave the parasitized cell, completing the virus life cycle. Often the host cell will be forced to give up its life, the victim of an elaborate ruse.

The keys to this ruse are the mechanisms used by DNA tumor viruses to activate dormant cells. One of the most interesting strategies is used by human papilloma virus (HPV), the agent that is present in the cells of more than 90 percent of cervical carcinomas. The close tie between HPV infection and this tumor seems to be more than coincidence. The epidemiology of this tumor has long suggested that it is a communicable disease: The larger the number of sexual partners a woman has had, the greater is the risk of cervical carcinoma. HPV infection undoubtedly plays a direct role in causing this cancer.

There are dozens of types of HPV, some of which cause common skin warts. Several HPV types grow well in the epithelial cell layers lining the cervix. On occasion, cervical carcinomas appear in tissues that have suffered chronic, decades-long infection by HPV. Still, the vast majority of women infected by the virus never go on to develop this malignancy. While HPV infection may be necessary to trigger most cervical carcinomas, it is clearly not sufficient by itself. In addition to the initial viral infection, other low-probability events must occur in the cervical epithelial cells before they become overtly cancerous.

HPV uses its *E7* viral oncogene to induce infected cervical epithelial cells to begin growing. The *E7* gene makes a protein product that interferes directly with the host cell's growth-controlling machinery by inhibiting the vital retinoblastoma protein. This disables the key braking apparatus used by the host cell to shut down its cell cycle clock and hence its own growth. The infected cell is now released from this restraint and moves into an active growth state that makes it a very good host for viral growth.

As described earlier, cells often respond to viral infection by committing apoptotic death. Deactivation of a cell's retinoblastoma protein can also cause it to enter apoptosis. This response, often rapidly mobilized, threatens the growth program of HPV by depriving it of a viable host in which to multiply. So HPV manufactures a second protein, $E6$, which disables the host cell's $p53$ protein, thereby blocking the apoptotic response. Having deactivated both the retinoblastoma and the $p53$ tumor suppressor proteins, HPV has cleared two major obstacles from its path to unconstrained multiplication inside the infected cell.

Other DNA tumor viruses have converged on the same solution for expediting their own growth. A monkey virus called SV40 makes a single oncoprotein, known as T antigen, which can bind both the $p53$ and retinoblastoma proteins of the infected cell and sequester them. Certain strains of human adenovirus are particularly interesting because of their variable effects; they cause common upper respiratory infections in their natural host species—humans—and tumors in unnatural hosts such as hamsters and rats. These adenoviruses, like the other DNA tumor viruses, make oncoproteins that deactivate both the retinoblastoma and the $p53$ tumor suppressor proteins, once again liberating infected host cells from growth constraints and converting them into more hospitable environments for viral replication.

Adenovirus has developed an additional feature to block apoptosis of its host. It carries another oncogene that, like the cellular Bcl-2 oncogene, has the capacity of blocking the apoptotic response. This extra feature helps ensure that the host cell's life will be preserved long enough to allow the virus to complete its full cycle of growth and replication and spawn a large brood of progeny.

Perhaps because it is so adept at multiplying and killing infected human cells, adenovirus is not connected with any human malignancy. In rodents, however, adenovirus will begin a round of growth replication by inducing the growth of infected cells, as is its usual habit. But in these unnatural host cells, the virus is unable to follow through with a subsequent round of viral multiplication and cell killing. These infected rodent cells are left alive, coexisting with the potent growth-stimulating viral oncogene inside them. This explains their subsequent behavior as tumor cells.

Viral infections seem to be responsible for only a small proportion of the cancers in the West. But study of the life cycles of DNA tumor viruses, notably SV40 and adenovirus, has provided scientists over the past two decades with a window on the inner workings of the cell cycle clock—the machinery that seems to be disrupted in all types of human cancer.

CHAPTER 15

MANY OBSTACLES IN THE ROAD: THE EVOLUTION OF A TUMOR

About 40 percent of the current population of the United States will develop cancer at some point in their lives. Half of these people will be cured and the other half will eventually die of this disease. In the mid-1990s cancer claimed more than half a million lives every year in the United States alone. From one perspective, this number seems horrendously high. The human body seems terribly susceptible to malignant disease. But there is another perspective that is much more reassuring. One-third of these cancer deaths were due to the use of tobacco, largely cigarettes. One-tenth were due to colorectal cancer, largely due to diet, most likely one high in meat and animal fat. By following a low-fat, low-meat diet and avoiding tobacco, an American can cut his or her risk of dying from cancer nearly in half, resulting in a risk of about one in ten. Some epidemiologists believe that strict adherence to a low-fat, vegetarian diet that includes ample fresh fruits and vegetables would reduce the risk even further.

For many cancer biologists, a one-in-ten cancer death risk seems remarkably low. This more positive view is inspired by another statistic: During a lifetime of seventy and more years, a human body will produce about 10 million billion cells. On 10 million billion separate occasions, cells will go through their cycles of growth and division. Each of these divisions represents an opportunity for disaster; the complexity of the cell cycle provides plenty of opportunity for something to go awry.

These numbers, taken together, lead to an interesting insight: While ten humans leading virtuous lives will collectively experience 100 million billion cell divisions, cancer is likely to kill only one of them. One fatal malignancy per 100 million billion cell divisions does not seem so bad after all.

Throughout this book, we have encountered a number of explanations for this very reassuring outcome. The body places numerous obstacles in the path of cells intent on forming tumors. It is these obstacles that hold down fatal malignancies to a very low number.

These hurdles must be transcended one after another, forcing cells to pass through a complex multistep process before they succeed in becoming truly malignant. The obstacles appear in many guises. Prominent among them is the signaling circuitry of the cell, which is hard-wired to resist disruption following the activation of an oncogene or the deactivation of a tumor suppressor gene.

The detailed design of this circuitry explains why multiple genetic changes are required before a cell can be diverted from its normal growth patterns. The circuitry is configured to resist destabilization by single malfunctioning components. Thus, the activation of an oncogene or the deactivation of a tumor suppressor gene will often have only minimal effects on the proliferation of a cell.

Then there are other bumps in the road. Should oncogenes succeed in transforming a cell, it may respond by triggering its apoptotic suicide program, depriving the oncogenes of their victory. And if apoptosis is avoided through one or another trick, then cell senescence and crisis threaten; only on the rare occasion that a cell passes through crisis and breaches the mortality barrier will it and its descendants have any chance of becoming life-threatening.

Even then, still other obstacles may lie in the path. Many researchers believe that the immune system erects a line of defense against developing tumors. For example, one class of white blood cell, termed natural killer (NK) cells, seems specialized to recognize and annihilate transformed cells. While this cell killing is apparent when NK and cancer cells encounter one another in a petri dish, the role of these NK warriors in preventing tumors in living tissues is still unproven. The potential anticancer role of NK cells remains an area of active research.

These multiple obstacles to malignancy force evolving precancerous cells to develop multiple genetic changes, each designed to circumvent or override one or another barrier. Each of these mutations, often affecting a proto-oncogene or tumor suppressor gene, is an infrequent event. Cancer is usually held at bay because it depends on a convergence of rare events that are unlikely to occur in an average human life span.

A DRIVE FOR FRESH BLOOD

If a small cohort of tumor cells overcomes all these hurdles, yet other difficulties loom. Like all cells throughout the body, the cells forming this incipient tumor require constant nourishment and oxygen. At the same time, they must continuously rid themselves of carbon dioxide and the waste products of their metabolism.

As long as a nest of tumor cells remains small—less than a millimeter in diameter—it can depend on diffusion to solve its logistical problems of supply and elimination. Molecules released by the cancer cell or its normal neighbors can diffuse over this short distance quite effectively. But once the clump of cells reaches the one millimeter size, it bumps up against a limit, a glass ceiling. Now the process of diffusion can no longer provide an adequate flow of nutrients and oxygen and a rapid removal of wastes. Soon, cells within the clump become starved and begin to choke on their own wastes. As described earlier, such anoxic cells often die from *p53*-mediated apoptosis.

The death rate of these cells from asphyxiation and metabolic poisoning begins to approach the rate at which these cells can regenerate themselves. Any gains made through cell proliferation are neutralized by attrition, and so the size of this tumor cell clump remains constant. Some microscopic nests of tumor cells may remain in this static state for years, possibly even decades.

To become threatening, these nests of tumor cells must break out of their futile cycles of division followed by starvation and asphyxiation. Such escape demands that the cells within these nests become truly creative: They must invent a better way of accessing nutrients and voiding wastes.

Their solution lies in developing their own blood circulation system. While the small band of tumor cells has survived half-starved, its normal neighbors nearby have all along enjoyed a reliable supply of nourishment and oxygen because of their close connections with the body's circulatory system. Unlike the small nest of tumor cells, normal tissue is interlaced with a dense network of capillaries. Often the array of capillaries is so dense that every cell in a tissue has direct access to an adjacent capillary. These small vessels, just wide enough to allow red blood cells to pass in single file, supply all metabolically active tissues throughout the body and carry off their wastes.

The capillaries are themselves constructed from cells. Such endothelial cells are highly specialized contortion artists that can flatten out and then bend themselves into tubes. The end-to-end joining of these tubular cells forms capillaries. Cells in normal tissues ensure the continued presence of capillaries by releasing specific growth factors that encourage the endothelial cells to stay on the job. In the event that certain cells become oxygen-starved, they will release VEGF (vascular endothelial growth factor) to induce endothelial cell proliferation and the formation of new capillaries. Without such encouragement, the endothelial cells will never thread their way through the interstices of a tissue to form dense, interwoven vessel networks.

To grow beyond the one-millimeter limit, a nest of cancer cells must invent a way to recruit capillaries into its midst. Judah Folkman, a surgeon in Boston, has spent the past two decades uncovering the cancer cells' strategy. Some cells in the clump, aping the normal cells around them, acquire the ability to secrete growth factors that attract endothelial cells from nearby tissue and induce these cells to multiply. Capillaries grow into the clump of cancer cells. Finally, the tumor cell clump has acquired direct access to oxygen- and nutrient-rich blood. Now this nest of cells can take off. Their proliferative agenda, frustrated for so many years, can now be fulfilled. Their numbers begin to increase explosively.

The growth factors released by these cancer cells are often called "angiogenic factors" because they encourage angiogenesis, the construction of blood vessels. These factors include VEGF and bFGF (basic fibroblast growth factor). The eventual success of the nest of tumor cells is closely tied to its ability to induce angiogenesis. Should members of this clump begin to elaborate high levels of angiogenic factors, their descendants, months later, will form tumors that are densely interlaced with capillaries; these cancers are often destined to grow aggressively and spread widely. Tumors that have poorly developed capillary networks are more indolent, and patients carrying such tumors usually have a much better prognosis. Indeed, some physicians use the presence or absence of dense capillary networks in tumor samples to determine the stage of tumor development and to predict its future course.

It remains unclear exactly how tumor cells acquire the ability to generate these in-growing blood vessels. Presumably some gene mutations in these cells trigger the sudden outpouring of angiogenic factors, which then paves the way for the tumor's long-term expansion.

BRANCHING OUT
....

A tumor mass one centimeter in diameter may contain as many as a billion cells. At first glance, the number seems huge, but it pales next to the number of cells in the body as whole—more than ten thousand times more. So a cancer this size is rarely life-threatening. In most places in the body, it probably will not compromise the functioning of a vital organ. Most tumors need to be far larger before they become lethal.

Of those patients who succumb to cancer, fewer than 10 percent die from tumors that continue to grow at the same site where they originally took root. In the great majority of cases, the killers are the metastases—colonies of cancer cells that have left the site of the original, primary tumor and have settled elsewhere in the body. It is these migrants, or rather the new tumors that they seed, that usually cause death.

The process of metastasis, which yields these far-flung colonies, is extraordinarily complex. To begin, cells within the primary tumor mass must breach the physical barriers that surround this growth. These barriers are most apparent in the case of carcinomas, which constitute the lion's share of human tumors. Carcinomas arise from epithelial cells, which form the linings of cavities in many internal organs and the outer layer of the skin. Epithelial cell layers are underlain by a structural meshwork of proteins, a "basement membrane" that separates the epithelium from connective tissue and circulation. This basement membrane is the first barrier confronted by carcinoma cells intent on leaving the mother tumor.

Because an intact basement membrane is usually impenetrable to cells, any invasion through the membrane requires that it be broken down. Invading tumor cells do this by releasing enzymes that cleave the proteins forming the basement membrane meshwork. Once this meshwork is dissolved, the tumor cells can gain access to the underlying tissue. Here too they may continue invasion and destruction by dissolving cells that represent physical obstacles to their forward march.

The invasiveness of the cancer cells depends on their ability to release proteases, enzymes specialized for cleaving protein chains into small fragments. Like angiogenesis, this secretion is a talent learned by tumor cells late in the multistep process of tumor progression. These protease enzymes are used by normal cells during the complex remodeling of tissue architecture that occurs when normal tissues are formed and repaired. As might be expected, the release and activity of these enzymes is usually under very tight control. Invading tumor cells subvert this control, then exploit and misuse these proteases; rather than releasing them in carefully measured dollops, they flood their surroundings with them.

High levels of these protease enzymes in a tumor tissue are readily detected in pathology laboratories. These laboratories often earn their keep by forecasting the future behavior of a tumor, a snippet of which has been provided them by the cancer surgeon. Like the presence of dense capillary beds, high levels of proteases in a tumor sample do not augur well for the patient. They signal the presence of cells that, through trial and error, have learned the art of destroying nearby tissue and are likely, for that reason, to spread far and wide.

The first small steps of invasion through the basement membrane result in minimal expansion of the tumor mass; in the case of an epithelial tumor, this localized mass is called a carcinoma in situ. But this invasion does place the invading cells in close proximity with vessels that provide highways to distant sites. Some tumor cells may use blood vessels as migration routes; others prefer to use the ducts of the lymphatic system. In either case, individual cells or small clumps of cells may break off from the mother tumor mass, raft through these vessels, and end up in distant locations.

These adventurers face an almost certain death. They must survive the swim through the circulation system, to which they are not adapted. They must cling to the wall of a lymph or blood vessel, force aside the endothelial cells lining the vessel, and then burrow through the sheathing around the vessel into underlying tissue. Once inside, these pioneer cells must find a way to thrive in an environment that is in many respects totally foreign.

Metastasizing colon cancer cells often found settlements in the liver. Breast cancer cells frequently find their way to bone. Lung cancer cells may metastasize to the brain. Each of these new sites presents a major challenge to the settler cells, which confront growth factors and physical structures to which they are unaccustomed.

By this time, late in tumor progression, the genomes of tumor cells have become quite unstable. This instability results in great genetic variability in the tumor cell population as a whole. New combinations of mutant genes are constantly being generated and tried out. As with Darwinian evolution, the rare cell that happens to carry particularly advantageous genes wins the race. Late in tumor development, the mutant genes conferring invasive or metastatic ability are at a premium.

Only very few tumor cells will acquire such genes through random accidents of mutation. The great majority will be ill suited for the rigors of metastatic voyage and settlement in new terrains, so their attempts to colonize distant sites will end up as suicide missions. By now, the primary tumor mass may have grown quite large and can afford to dispatch a large, continuous stream of scouts on these missions. Even a seemingly impossible mission will succeed if tried often enough, so some new colonies will be founded and then thrive at distant sites. Sooner or later, these metastases begin to compromise the functioning of the host tissues in which they have taken root. Only then is the cancer patient placed at death's door.

The mechanics of all these steps are still poorly understood. As they are swept through lymph or blood vessels, metastasizing cells display on their surfaces certain receptor molecules that enable them to tether themselves to the walls of the vessels. There are many kinds of such anchoring receptors. Each different type of anchor line enables the metastasizing cell to bind itself to a distinct molecular homing site. The number and complexity of these anchoring receptors has impeded progress in understanding how any one of them operates.

Our understanding of metastasis is still fragmentary. The principles that guide the migratory routes of most cancer cells are, for the moment, as mysterious as those that guide the monarch butterfly. For the cancer researcher, the process of metastasis remains a terra incognita, still largely unexplored.

CHAPTER 16

AN END TO THE SCOURGE: USING THE KNOWLEDGE OF CANCER'S CAUSES TO DEVELOP NEW CURES

The scientific revolution that began two decades ago continues to this day. We have learned much about the invisible forces that create human cancer. Knowing the causes of many tumors, we should be able to prevent their appearance, or if they appear, to treat them and achieve permanent cures. The powers of genetics and molecular biology have been brought to bear on the most complex of human diseases. While not all the pieces in the puzzle are in place, the outlines of a complete explanation of cancer are in clear view. The game plan of modern biomedical research has been validated once again: Take apart complex problems piece by piece, reduce them to simple and analyzeable elements, and then derive clear, unassailable truths. In two short decades, we have moved from contentious debates about cancer's origins to widely accepted, detailed descriptions of the forces behind human malignancies.

The major insight afforded by this revolution is that cancer is a disease of damaged genes. We now know the identities of many of the culprit genes—oncogenes and tumor suppressor genes. They control the behavior of the cells in which they operate; the cells respond in turn by generating tumors. To be sure, many cancer-related genes still remain to be identified and isolated by gene cloning. The means by which many genes influence cell behavior also remain to be discovered.

We know that the agents that provoke human cancer promote, directly or indirectly, the creation of mutant genes. We know that the appearance of a human tumor requires a succession of mutations, each responsible for perturbing a distinct cellular growth-controlling gene. And we know that processes that threaten the integrity of the cell genome, including defects in its maintenance and repair, strongly influence the rate at which cancers appear.

The discovery of various growth-controlling genes has provided us with a view of the complex decision-making circuitry that lies within each human cell. Biologists have cataloged the diverse behaviors of cells for more than a century. Cell behavior seemed to have its own logic, determined by submicroscopic vital forces hidden deep within. We now understand that logic in terms of critical signal-processing proteins that determine the cell's responses to a wide variety of stimuli; these proteins assemble to form an elaborate signal-processing circuitry. Every week new pieces are added to the wiring diagram of this circuitry. Its design—its interconnections and the actions of its component parts—determines how cells behave.

Knowledge of this circuitry will provide the ultimate answers for those interested in understanding cancer. There are no deeper or more subtle mechanisms hiding in some secret corner of the cell. The answers are all there, or will be found there shortly. Two decades ago, we knew nothing of all this.

This research has allowed us to discover the heart and mind of the human cell, the cell cycle clock. As master of the cell's fate, it governs decisions to grow or differentiate. While research on this clock is still young, we already know that this clock suffers damage in most and probably all types of human tumors. Here again, the outlines of a definitive description are in place, but many critical details still need to be fleshed out.

Many of the genes discovered over the past decade have also afforded us a bridge between cancers that are preordained from the moment of conception—the familial cancer syndromes—and those that strike sporadically, the results of random genetic accidents occurring during an individual's lifetime. These two classes of cancer are not distinct diseases, as some imagined, but rather manifestations of a common gene repertoire that may suffer damage either before or after the sperm penetrates the egg.

CONQUERING CANCER BY PREVENTING IT

What are the prospects that the rich harvest of information described here will have palpable effects on the death rate from cancer? Simple logic would dictate that the cure of a disease is most likely to come from understanding its causes. Hence our recently gained knowledge of genes and proteins should take us far in conquering this disease. But cancer's ultimate causes really begin far outside the individual cell, in our environment, in the food we ingest and the smoke we inhale. We must address these ultimate roots of cancer before we make substantial reductions in cancer incidence. Genes and proteins will help us little here.

The precedent set by other major diseases over the past two centuries is clear: Reductions in mortality have come from improvements in personal sanitation, nutrition, access to clean water, and immunization. By extension, the big decreases in cancer deaths will likewise come from preventing disease rather than discovering new cures.

Major reductions in cancer mortality will derive from identifying and eliminating discrete causes of the disease—in particular, certain aspects of diet and lifestyle. Much of this job is the purview of the epidemiologists. Indeed, we have already learned much from them. They have framed the problem, defined its scope, breadth, and depth. They also have disabused us of a couple of notions widespread in some circles: that the industrialized West is being inundated with a cancer epidemic, and that most of this inundation is traceable to chemical pollutants in the air and in the food chain.

Except for breast cancer and tobacco-related cancers, the rates of most kinds of cancers have held steady over half a century, a time when environmental pollution has increased substantially. At most, only several percent of human cancer can be ascribed to man-made environmental agents. In 1930, the annual rate of mortality from cancer in the United States was 143 per hundred thousand population. By 1990, the rate had increased to 190 per hundred thousand. These numbers are adjusted for changes in the age distribution of the population and for the fact that, as mentioned repeatedly here, cancer incidence is a strongly age-dependent disease.

Almost all of this age-adjusted increase in cancer deaths flows directly from tobacco consumption. During the 1990s, one-third of cancer deaths in the United States were caused by tobacco. The obvious response—reduction in tobacco usage—has already shown its effectiveness: By 1990, the century-long increase in lung cancer death rates in men was reversed. Without the contribution of lung cancer, the overall age-adjusted death rate from cancer would have declined 14 percent between 1950 and 1990.

Breast cancer death rates have increased modestly—by about 10 percent between 1960 and 1990. The incidence of the disease appears to be growing more rapidly, but a greater proportion of cases are being cured, largely through surgery.

The causes are hotly debated. One view, rapidly gaining ground, ascribes the bulk of this increase to modern nutrition and reproductive habits, which have collaborated to increase the number of menstrual cycles that women experience during their lifetimes. An additional factor, poorly understood, is the protective effect of early pregnancy, which seems to place the breast tissue in a state that resists cancer onset later in life. Delayed child rearing has therefore contributed to the climbing incidence.

Diet plays a critical role in perhaps half of all human cancers, but most of the cancer-causing components in the food chain are hard to identify. The Western diet results in a colon cancer rate ten to twenty times higher than that in parts of central Africa. One likely culprit is the high content of meat and animal fat common in the Western diet. The cooking process may play a substantial role here; potent carcinogens are generated when meat, especially red meat, is heated to high temperatures. Asia is also plagued by high rates of diet-related cancers. The Japanese diet results in a stomach cancer rate as much as six times higher than in the United States. Salted, fermented, and smoked foods have been indicted as likely causes.

Plant foods present a double-edged sword, since they contain both cancer-causing and cancer-preventing compounds. Vegetables provide vitamins such as A, C, and E, which neutralize some important carcinogens, notably the oxidants and free radicals generated during normal cell metabolism. On the other hand, some plant components may actively contribute to cancer. Plants have evolved elaborate chemical defense systems to make themselves unpalatable to insect predators. Among these are compounds that are registered as potent mutagens in the Ames test. Ames himself has cited the case of a new strain of celery that was bred to require less synthetic insecticide during its cultivation; the increased insect resistance shown by this celery correlated with a tenfold increase in a potent mutagen that is found naturally in this vegetable.

Like all other plants, celery carries dozens of carcinogenic and anticarcinogenic constituents. And celery is only one of dozens of plants in a normal diet. Each contributes its own spectrum of simple and complex organic compounds to our food. The interactions between these mixtures of natural compounds and their effects on human metabolism are complex beyond measure. It will take decades to determine which natural common foodstuffs make us healthy and which ones shorten our lives.

This complexity notwithstanding, some conclusions have already begun to emerge. As mentioned earlier, it seems that nearly half of current cancer could be prevented through avoidance of tobacco use and high-fat, high-meat diets. But what of the remaining half? Cancer will undoubtedly be with us for many generations to come, striking even those who live the most virtuous of lives. How will we deal with these seemingly unpreventable tumors?

EXPLOITING GENES AND PROTEINS TO DEAL WITH CANCER

Even after we have identified all of the external causes of cancer, human behavior will never be fully responsive to the discoveries of epidemiologists. Even more important, the extraordinary complexity of the human body dictates cancer's inevitability. All complex machinery breaks down sooner or later. Given enough time, cancer will strike every human body. It is here that the recently discovered genes and proteins will help us. They will allow us to deal with the cancers that we cannot avoid.

Early detection of tumors will prove increasingly important. The discovery and removal of a tumor mass in its initial stages often results in a cure. But early detection presents two major difficulties. First, nests of cancer cells must be discovered while they are still very small. As we saw earlier, a tumor of one centimeter diameter constitutes less than 0.01 percent of the body's mass. Few currently available biochemical assays are sensitive enough to detect such minute entities.

Second, cancer cells, especially those in the early stages of tumor progression, are in almost all respects very much like normal cells. The task of finding distinctive markers that are specific to cancer cells is daunting. Almost all proteins that have been proclaimed "tumor-specific" have later been found to be produced also by normal tissues somewhere in the body.

In spite of these failures, the most attractive approaches to tumor detection still derive from identifying genes and proteins that are unique to cancer cells. Mutant oncogenes, tumor suppressor genes, and their protein products come to mind. The mutant *ras* oncogene present in about one-quarter of all human tumors carries a DNA sequence not found in normal cells; and this gene causes the production of a *ras* protein having a unique, unnatural structure.

Knowing this, some researchers have attempted to find cells bearing mutant *ras* oncogenes in the colon. The job is made much easier by the fact that tumor cells, like their normal counterparts, are shed continually in large numbers into the fecal stream. Using an ultrasensitive DNA analytic technique, David Sidransky at Johns Hopkins University has detected mutant *ras* genes in the DNA in stool samples. These analyses were done on samples from individuals bearing colonic tumors that had already been identified by other means. This technology must be improved and made even more sensitive, but its long-term promise is already clear: It should allow the discovery of colonic tumors early in the course of their progression to malignancy, while they are still treatable through surgery.

The same strategy may eventually be applicable to tumors of other hollow organs, including the bladder, uterus, and lung. In each case, cells are shed into the cavities of these organs. Shed bladder cells can be found in the urine, while shed lung cells can be detected in the mucus high in the bronchial tubes. As in the colon, analysis of shed cells offers the prospect of early detection and increased chances of cure.

Familial tumors also contribute to a substantial fraction of human cancer incidence. Some researchers estimate that as much as 10 percent of all human cancer derives from inherited genes. The ability to predict inborn susceptibility to cancer thus represents another highly useful dimension of early detection.

For colon cancers, the familial polyposis and HNPCC syndromes may together account for more than 10 percent of all cases. A comparable proportion of breast cancers are likely to be associated with inheritance of mutant alleles of the *BRCA*1 and *BRCA*2 genes. Over the coming decade, a proportion of almost every type of common cancer will be found to be caused by inherited mutant versions of identified genes, often tumor suppressor genes.

The technology for detecting mutant genes in small tissue samples is improving rapidly. Soon, a mutant inherited gene predisposing an individual to one or another cancer will be readily detectable using only several drops of that person's blood. Similar analyses will be available for prenatal diagnosis, to be used in families known to be afflicted with unusually high rates of certain cancers. These tests will identify those members of a family who are at high risk and those who are spared that risk. Family members who are found to be at risk will need to be monitored throughout their lives. In the case of especially life-threatening conditions, including familial polyposis and breast cancer, the patient may decide to have the target organ removed before a malignant tumor appears.

Still, these powerful genetic techniques will provide incomplete protection. Thorough, population-wide screening for inborn cancer susceptibility will remain economically and logistically impossible, and most small nests of sporadic tumors will continue to elude the nets cast by those using state-of-the-art detection techniques. For these reasons, we will continue to confront many tumors that will be diagnosed only after they have become large and symptomatic. Then, as now, the powers and limitations of antitumor therapy will determine life or death. Over the past decade, the long-term survival rates of patients carrying many types of solid tumors have remained rather constant. We need to develop radically new kinds of therapies to make further improvement.

The basic molecular research described in this book promises large payoffs in this respect, although these payoffs will only come years in the future. In the course of understanding the defective wiring diagram of cancer cells, researchers have uncovered many genes and proteins that are attractive targets for a new generation of antitumor drugs.

The beginnings of this new wave of drug development are already with us. Compounds under development by drug companies show great potency in their ability to block the cell's ability to manufacture functional *ras* protein. Most surprising are the specificities of these drugs, which strongly affect tumor cell growth but have relatively small effects on normal cells, even though the latter cells are known to depend on the normal version of the *ras* protein for their growth and survival.

Monoclonal antibodies may also prove useful. These antibodies, made in mice, bind specifically to certain human proteins, ignoring all others. They are infallible homing devices. Some have been created that are specifically reactive with the cell surface receptors—the EGF and *erb* B2/*neu* proteins—that are displayed in unusually high concentrations on the surfaces of breast cancer cells.

These antibodies can be exploited in two ways. First, radioactive atoms may be linked to them. When injected into a patient, they home to tumor cells expressing large amounts of one or another of these receptors, concentrating radioactivity in the region of the tumor. This radioactivity can be scanned by computerized imaging devices, making it possible to visualize a tumor that would not otherwise stand out using standard imaging techniques such as CAT scans. Second, toxins can be linked to the antibodies. This converts the antibodies into "smart bombs" that guide the toxins to the tumor cell targets.

While attractive in concept, both applications of monoclonal antibodies are complicated by the fact that normal cells may also express these receptor target proteins, though at lower levels. For that reason, a toxin-linked antibody may inadvertently destroy normal cells that happen to display some of the receptor molecules targeted by the antibody. Use of radio-labeled antibodies may succeed in revealing the outlines of a tumor, but again, normal cells expressing the target antigen may diffuse the image to the point that it is no longer useful to the surgeon seeking to localize the tumor.

The greatest innovation in cancer chemotherapy may come from the recent realization of the importance of apoptosis. Many chemotherapeutic drugs succeed by inducing this programmed suicide of tumor cells. Because the *p53* protein renders many cells susceptible to apoptosis-inducing drugs, the oncologist of the future may ascertain the genetic status of the *p53* gene in a tumor before planning a chemotherapeutic strategy.

The majority of tumors lack normal *p53* function and so are likely to be less responsive to currently used chemotherapies. Researchers intent on developing new anticancer strategies will need to address these tumors, perhaps by developing ways of triggering apoptosis even in the absence of functional *p53* protein. This directs attention to the cellular circuitry that controls the apoptotic response. The *bcl*-2 proto-oncogene is one of a dozen or more genes regulating this important response. The role of many of these—as either promoters or antagonists of apoptosis—is still being discovered. Once we understand the logic of this circuitry, we will find new ways of inducing cells, including cancer cells, to commit suicide. The prospects for the development of totally novel anticancer therapeutics are bright!

A ROAD AHEAD
....

By the end of the first decade of the new century, the elements of the cellular wiring diagram will be known in exhaustive detail. Every signal-transducing protein will have been put in its place on a large chart describing how cells receive and process the signals that influence their growth and differentiation.

By then, a new set of talents will be brought to bear on the cancer problem. Mathematicians with expertise in analyzing complex multicomponent systems will explain to biologists how the minicomputers inside cells actually function. They will tell us how the mind of the cell works, and how it becomes derailed during tumor progression.

Until recently, the strategies used to find the genes and proteins that control the life of the cell have depended on ad hoc solutions to formidable experimental problems, cobbled together by biologists who lacked better alternatives. Time and again, serendipitous discoveries have allowed new pieces to be placed in the large puzzle. But sustained, steady advances have been elusive, causing most researchers to cast about more or less aimlessly, following up tantalizing clues, few of which pan out. The substantial progress that has been made has come only because hundreds of independent research groups have each contributed piecemeal advances. Because of their collective efforts, the information harvest has been bountiful.

Soon, all this will change dramatically. In the years to come, we will begin to learn in a more systematic way how the cell is put together. Many of the new advances will be driven by the Human Genome Project, the worldwide cooperative undertaking to catalog the entire gene library carried in human cells. Soon we will know whether the human genome carries eighty thousand or one hundred thousand genes. The base sequences of these genes will often provide broad hints as to their roles in the life of the cell.

Until recently, the search for tumor suppressor genes, including those that create inborn susceptibility to cancer, has been a slow and agonizing process. The techniques used have been imprecise and always labor intensive. Discoveries of critically important genes often have depended on little more than dumb luck. Once the human genome is known in intimate detail, the catalog of tumor suppressor genes will explode. Within a decade's time, we will have identified almost all of them and will understand their roles in most kinds of human cancers.

Other technologies will also be brought to bear. All humans carry genes that affect their susceptibilities to various types of cancer. In the great majority of cases, these genes act in subtle ways, influencing the ways in which we detoxify potential carcinogens, the efficiency with which we keep our DNA in order, or how effectively we kill off errant cells on their way to becoming cancerous. Because humans are a genetically heterogeneous species, each of us carries a different combination of these genes. For that reason, the appearance of any human cancer represents the confluence of random events and the interplay of a large cohort of variable genes.

At present, cancer geneticists focus their energies on understanding the roles of single genes and how each of these influences the creation of a tumor. But the great majority of tumors arise from the combined actions of a constellation of genes, not single ones operating in isolation. In the future, new kinds of mathematics will make it possible to understand the origins of polygenic cancers, in which cohorts of genes act in combinatorial ways to favor the appearance of cancers. In ten to fifteen years, we should be able to predict with some precision an individual's risks for contracting a whole variety of polygenic cancers. These predictions will be made quickly and cheaply, exploiting the great improvements coming in both data processing and the automation of DNA sequence analysis.

The repertoire of genes laid out by the gene mappers will hardly provide all the answers. At present, the three-dimensional structures of most proteins cannot be predicted from the DNA sequences that specify them. This problem will surely be solved by the first decade of the new century. With that solution will come the ability to predict how many proteins involved in the cancer process operate, even without direct biochemical analysis of the proteins themselves.

In spite of these sea changes in information processing and analysis, the hands-on work of biochemists and geneticists still will be central. They will need to figure out how individual proteins talk to one another inside the living cell. Here too, technology will create a revolution. Powerful gene-cloning strategies, some already in place, will reveal which partner proteins interact with one another inside the soup of the living cell, and how these protein-to-protein interactions generate the large communication networks by which cells decide to grow, differentiate, or die.

Finally, the ways in which new anticancer drugs are discovered will change radically. Until recently, vast libraries of candidate drugs assembled from hundreds of thousands of distinct chemical compounds have been screened for those few that have potent antitumor effects. These searches have been laborious almost beyond measure. And they have not been guided by knowledge of the molecular mechanisms that underlie cancer formation.

Drug development will change in two dramatic ways. The screening will be increasingly relegated to robots—a trend that has already begun. And specific proteins inside the cancer cell will be chosen as targets for attack by newly developed drugs. The structure of these proteins will serve as a guide to the chemists intent on designing drugs to attack their function. The ability to knock out a critical cancer-causing protein will no longer be hit or miss.

The new "rational drug design" will lead quickly to potent inhibitors. With increasingly detailed information on the metabolisms of normal and cancerous cells, it will become possible to design highly selective drugs that strike at cancer cell targets while leaving the cells of normal tissues relatively unaffected. The horribly uncomfortable side effects of cancer therapy will be mitigated, perhaps even eliminated.

Those who engineer these successes will view the discoveries of the last quarter of the twentieth century as little more than historical curiosities. But for the moment, we allow ourselves to see things quite differently. We can glory in having lived while the foundations were being laid. We have moved from substantial ignorance to deep insight. We have lived through times of great excitement!